Project Contacts

I0476904

EPA Project Leads

Stephanie Bertaina
Office of Sustainable Communities
U.S. Environmental Protection Agency
1200 Pennsylvania Ave., NW (MC 1807T)
Washington, DC 20460
Tel 202-566-0157
bertaina.stephanie@epa.gov

Rosemary Monahan
Office of the Regional Administrator
Region 1
U.S. Environmental Protection Agency
5 Post Office Square (ORA 18-1)
Boston, MA 02109-3912
Tel 617-918-1087
monahan.rosemary@epa.gov

FEMA Project Lead

Marilyn Hilliard
Federal Emergency Management Agency
Region 1
Mitigation Division
99 High St., 6th Fl.
Boston, MA 02110
Tel 617-956-7536
marilyn.hilliard@fema.dhs.gov

Vermont Project Lead

Faith Ingulsrud
Vermont Department of Housing & Community
Development
Community Planning & Revitalization Division
National Life Building 6th Floor
1 National Life Drive
Montpelier, VT 05620-0501
Tel 802-828-5228
faith.ingulsrud@state.vt.us

Contractor Leads

Kate Marshall, Project Manager
SRA International, Inc.
3434 Washington Blvd.
Arlington, VA 22206
Tel 703-284-6234
kate_marshall@sra.com

Gavin Smith, Principal Researcher
University of North Carolina at Chapel Hill
Department of Homeland Security
Coastal Hazards Center of Excellence
100 Europa Drive
Suite 540, CB 7581
Chapel Hill, NC 27517
Tel 919-445-9395
gavin_smith@unc.edu

Cover Photo: Mad River, Vermont.
Credit: EPA.

TABLE OF CONTENTS

Executive Summary

Flooding from extreme storm events has affected many communities across the country, causing billions of dollars of damage annually. Moreover, climate change projections suggest that storms will likely become more powerful in many regions of the country in the future. In light of these trends, many communities are recognizing the need to improve disaster recovery and long-term flood resilience planning.

Communities throughout Vermont faced this reality when Tropical Storm Irene hit in 2011, devastating infrastructure, communities, and lives. In 2012, in the wake of Irene, the state of Vermont requested technical assistance from the U.S. Environmental Protection Agency (EPA) and the Federal Emergency Management Agency (FEMA). The assistance focused on incorporating smart growth principles into state policies, local development regulations, and Hazard Mitigation Plans to increase community flood resilience. "Flood resilience" means measures taken to reduce the vulnerability of communities to damages from flooding and to support long-term recovery after an extreme flood.

Smart growth and more environmentally and economically sustainable approaches to development can help communities become more resilient to future flooding by protecting vulnerable undeveloped lands, siting development in safer locations, and designing development so it is less likely to be damaged in a flood. Communities that identify areas that are safer for development and then implement smart growth approaches in those areas will be most successful at creating more flood-resilient places. EPA's assistance provided options for communities and the state to consider as they work to recover, rebuild, and plan for a more resilient future.

Communities can take some initial steps to enhance their flood resilience:

- They can update and integrate their community or comprehensive land use plans with Hazard Mitigation Plans, ensuring that the comprehensive plan identifies future growth areas in safer locations and that hazard mitigation activities are consistent with the comprehensive plan priorities. If these plans are not coordinated, they might inadvertently act at cross-purposes.
- They can conduct an audit of policies, regulations, and budgets to ensure consistency with flood resilience goals outlined in their community plans and Hazard Mitigation Plans.
- They can amend existing policies, regulations, and budgets or create new ones that help achieve the flood resilience goals outlined in their plans.

Specific local land use policy options to improve flood resilience are organized into four categories, representing different geographic areas in a community:

- River Corridors[i]: Conserve land and discourage development in particularly vulnerable areas along river corridors such as flood plains and wetlands.
- Vulnerable Settlements: Where development already exists in vulnerable areas, protect people, buildings, and facilities to reduce future flooding risk.
- Safer Areas: Plan for and encourage new development in areas that are less vulnerable to future floods.
- The Whole Watershed: Implement enhanced stormwater management techniques to slow, spread, and infiltrate floodwater.

---

[i] "River Corridors" are areas of land that include the river channel and adjacent lands needed for the river to adjust laterally over time and still maintain its natural stable form. The surrounding areas of land may be developed or undeveloped.

The policy options in these categories offer multiple and interrelated benefits. For example, directing development out of flood plains not only keeps people and property safe, it also protects the ability of flood plains to hold and slow down flood water before it reaches downstream settlements.

State-level policies also can support flood recovery and local flood resilience planning efforts. State agencies can partner together to:

- Audit all state programs to determine how well they achieve flood resilience goals.
- Develop a comprehensive recovery plan before the next flood happens.
- Develop a personnel plan that delineates who will assist with post-disaster recovery.

Individual state agencies that manage natural resources, environmental protection, transportation, emergency management, commerce, community development, economic development, housing, and agriculture can also make changes to their policies and programs to ensure that they are helping communities become more resilient to future floods.

While land use decisions that affect a community's flood resilience might seem to happen incrementally or opportunistically, they are often guided by plans, policies, and regulations that shape development over time. Vermont's experience with Tropical Storm Irene suggests that coordinating local and state agency policies, plans, and actions can help facilitate disaster recovery and promote safer growth.

The Flood Resilience Checklist (in Appendix C) and the land use policies, regulations, and strategies outlined in this report (many of which are listed in Appendix D) can help communities enhance their flood resilience. Ultimately though, it is up to the state and communities to select the appropriate flood resilience policies, adjust them to meet their specific contexts, and allocate resources accordingly. Each jurisdiction can weigh its resilience goals with other community priorities and determine the best policies and approaches that will help the community meet its objectives.

1. Introduction

A. Background

Many communities across the United States have experienced damage from flooding. Despite the use of expensive, engineered solutions to reduce flooding risk, such as elevating buildings and constructing levees, flood damage losses in the United States continue to grow.[1] Moreover, climate change projections suggest that floods will intensify in most regions of the United States, especially in the Midwest and Northeast.[2] According to the National Climate Assessment,[ii] "the Northeast has experienced a greater increase in extreme precipitation over the past few decades than any other region in the United States; between 1958 and 2010, the Northeast saw a 74% percent increase in the amount of precipitation falling in very heavy events."[3] Rainfall in New England is expected to continue to increase due to climate change, a trend that will almost certainly increase the risk of river-related flooding in this part of the country in the future.

These trends are creating a sense of urgency among communities, particularly those in states like Vermont that are expected to experience increased flooding in the future, to look for better ways to deal with flooding and build flood resilience. Resilience generally refers to "a capability to anticipate, prepare for, respond to, and recover from significant multi-hazard threats with minimum damage to social well-being, the economy, and the environment."[4] This project focused specifically on resilience to flooding, including a community's capacity to plan for, respond to, and recover from floods.

The state of Vermont experienced widespread damage from river flooding as a result of Tropical Storm Irene in 2011. Irene damaged more than 500 miles of roadways and around 200 bridges (with estimated rebuilding costs of $175-250 million); released hazardous waste that contaminated floodwaters, sediment, and soil; breached municipal wastewater treatment plants; and caused agricultural losses by damaging barns and flooding crops.[5,6] The Mad River Valley—located in north central Vermont, west of Montpelier (see Figure 1)—was one of many regions in the state that was affected by Irene. Many historic structures, homes, and businesses in the Mad River Valley were flooded. Irene was particularly damaging to communities in Vermont, but communities throughout the state and region have experienced flood damage decade after decade, underscoring the need for improved hazard mitigation planning at the state, regional, and local levels.

Shortly after Irene, several Vermont state agencies and communities in the Mad River Valley requested technical assistance from the U.S. Environmental Protection Agency (EPA) and the Federal Emergency Management Agency (FEMA). The state and Mad River Valley communities sought help with incorporating smart growth and resilience approaches into their development plans, regulations, and Hazard Mitigation Plans to increase their flood resilience.

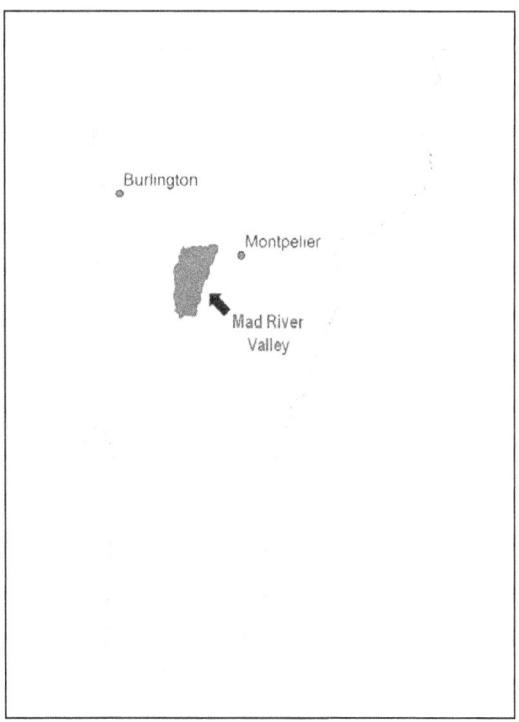

Figure 1. This project focused on the Mad River Valley in Vermont. Credit: EPA.

[ii] All relevant website links are spelled out in the endnotes.

Smart growth is development that is compact and walkable, provides a range of housing and transportation choices, and fosters distinctive, attractive communities with a strong sense of place.[7] Smart growth approaches use land efficiently, enhance community vitality, protect natural resources, reduce costs for public services, save taxpayers' money, and create a higher quality of life.

According to the Vermont Natural Resources Council, a nonprofit environmental organization that aims to protect Vermont's natural resources and environment, Vermont's distinctive sense of place is influenced by the state's landscape of compact cities and villages surrounded by working farms and forests. Smart growth approaches to development can help preserve Vermont's sense of place by promoting development that is good for the state's economy, community, and environment.[8]

However, smart growth approaches alone cannot completely address flooding risk. Communities that seek to become more resilient to future flooding must also protect vulnerable undeveloped lands, site development in safer locations, and design development so it is more resilient to floods. Communities that identify areas that are safer for development and then implement smart growth approaches in those areas will be most successful at creating more flood-resilient places.

EPA and FEMA provided assistance through the Smart Growth Implementation Assistance Program (see Appendix A for more information on the program), funding a team of national experts in hazard mitigation, flood recovery, land use planning, and state policy. The team reviewed state policies, local development regulations, community plans, and Hazard Mitigation Plans and developed policy options for the state and communities to consider. In October 2012, the team visited the Mad River Valley and presented initial policy options for the state and communities to consider. During the visit, the team solicited feedback from state and local leaders and the community about those ideas during a public meeting; and then refined the policy options outlined in this report. More information on the project is in Appendix B.

This project included two elements: an assessment of local policies and an assessment of state policies to enhance flood resilience. The local policy assessment, which was funded by EPA and completed by consultants from SRA International, Inc., Clarion Associates, and CSA Ocean Sciences, Inc., focused on two Mad River Valley communities, Waitsfield and Moretown, which are representative of towns within the Valley and throughout the state. The goal was to offer policy options to these communities to help them update and strengthen their policies and strategies to improve flood resilience and that other local governments in Vermont and elsewhere in the United States could also consider.

Figure 2. These images show the flood damage in the Mad River Valley from Tropical Storm Irene: a damaged home along Vermont Route 100 adjacent to Moretown Village (left) and a damaged building in Waitsfield (right). Credits: Lars Gange and Mansfield Heliflight (left), Jeff Knight, The Valley Reporter (right).

Faculty and staff from the University of North Carolina at Chapel Hill's Department of Homeland Security Coastal Hazards Center of Excellence (Coastal Hazards Center team) led the state policy assessment; FEMA funded the state policy assessment. The Coastal Hazards Center team focused on potential barriers to flood response, hazard mitigation, and disaster recovery at the state level, including the degree to which state programs and policies support or hinder local governments' ability to incorporate smart growth and flood-resilience measures into their day-to-day activities.

B. Community Context

The Mad River Valley, which lies about 15 miles west of the state capital of Montpelier, is a historic, scenic area that is home to two popular ski resorts, Sugarbush Resort and Mad River Glen. The rushing waters of the rocky Mad River cut through this deeply incised valley, attracting kayakers, canoeists, and anglers. There are five small towns in the watershed—Warren, Waitsfield, Fayston, and a portion of Moretown and Duxbury (see Figure 3).

The two municipalities involved in this project, Waitsfield and Moretown, each have populations of around 1,700. Both have grown faster than the state of Vermont as a whole over the past two decades, but their population growth rate has been less than 1 percent annually—a very modest pace. Both jurisdictions are typical of many smaller riverfront communities in Vermont. They have compact, historic village centers that are next to the Mad River in high-flood hazard areas.

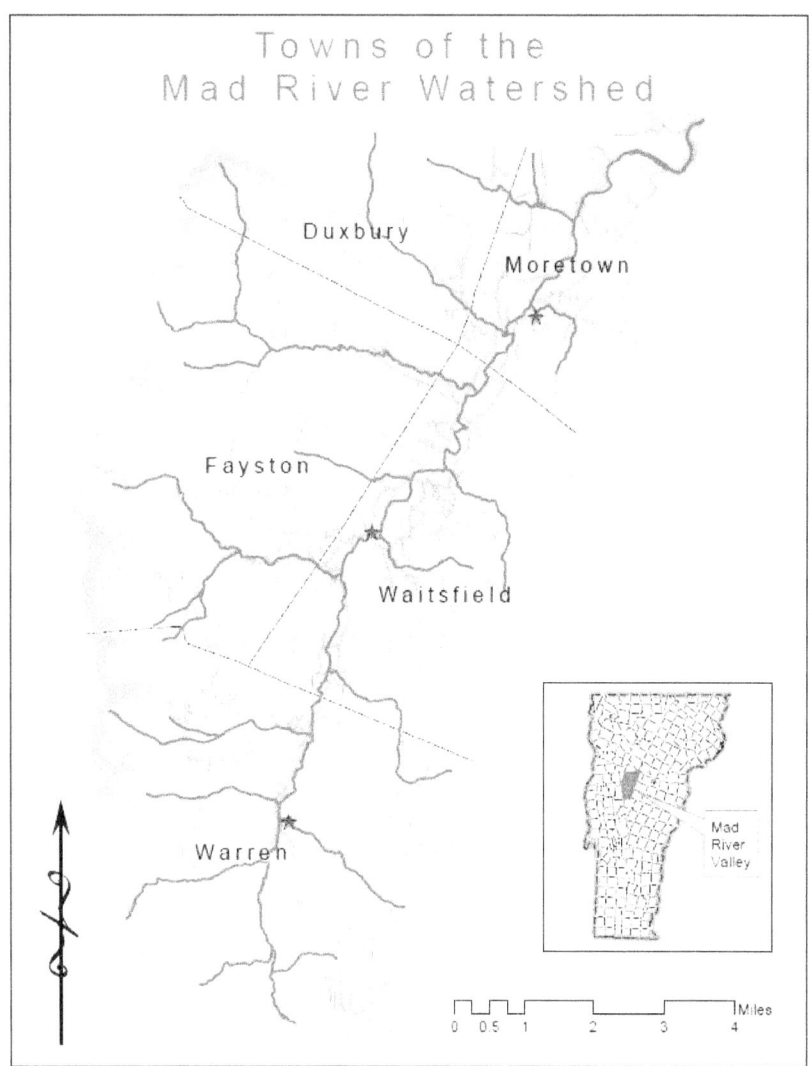

Figure 3. This project focused on Waitsfield and Moretown, two of the five communities in Vermont's Mad River Valley. Credit: Mad River Watershed Conservation Partnership.

Because Vermont has no county governments, the municipalities have land use planning and regulatory authority over the surrounding large tracts of forests and open space. Waitsfield, along with Fayston and Warren, participates in a sub-regional organization, the Mad River Valley Planning District, which provides planning support and inter-town coordination for the three towns, amplifying the planning capacity for those communities. Moretown does not participate in the Mad River Valley Planning District and has a very small, part-time staff to handle community planning issues.

The Mad River Valley jurisdictions have begun revamping their hazard resilience policies and strategies and have a solid foundation upon which to make additional changes. For example, Mad River Valley communities have access to critical data on the location and nature of fluvial (river-related) erosion hazards that can cause damage to public infrastructure, homes, businesses, and other private investments during flooding events.[9]

These data—available from organizations like the Vermont Agency of Natural Resources, Central Vermont Regional Planning Commission, and the Friends of the Mad River can help Mad River Valley communities determine where they can more safely locate development in the future.

Figure 4. Waitsfield, Vermont was one of two Mad River Valley communities that received technical assistance from EPA and FEMA. Credit: Clarion Associates.

Waitsfield completed an update to its town plan in 2012 and is considering amendments to its development codes and Hazard Mitigation Plan. In 2013, Moretown began the process of updating its town plan. This project aimed to help these communities identify smart growth and resilience approaches to development that they could incorporate into their plans and development regulations in their continued efforts to enhance their flood resilience. These approaches can also be considered by other communities throughout the state and country that are facing similar issues.

2. Overall Strategies for Flood Resilience and Disaster Recovery

Vermont communities that want to better withstand and recover from flood-related disasters in the future might wish to consider updating, integrating, and revising their plans, policies, and regulations to ensure that they are consistent with the community's resilience goals and objectives. These approaches, while specifically helpful for Vermont communities, might also be useful for other communities seeking to enhance their flood

> **Overall Strategies for Flood Resilience**
>
> This section of the report corresponds with the "Overall Strategies to Enhance Flood Resilience" section in the Flood Resilience Checklist in Appendix C. Please see the checklist for a list of strategies to consider to enhance overall flood resilience.

resilience. Several basic steps might help communities get started on their road to resilience:

A. Update and integrate comprehensive plans and Hazard Mitigation Plans.

B. Conduct thorough policy and regulatory audits.

C. Amend zoning, subdivision, and stormwater policies and regulations to match plans.

D. Consider participating in the National Flood Insurance Program Community Rating System.

A. Update and integrate comprehensive plans and Hazard Mitigation Plans.

Many local governments adopt comprehensive plans to guide future land use decisions in their communities. State governments and FEMA also encourage communities to prepare Hazard Mitigation Plans to improve planning for and reduce or eliminate risk from natural hazards.[10] A community must have a Hazard Mitigation Plan to receive Hazard Mitigation Grant Program funding from FEMA.

Comprehensive plans shape communities' flood resilience by determining where and how development will be built in the future, and Hazard Mitigation Plans shape communities' flood resilience by informing how communities will plan for and reduce or eliminate risk from natural hazards such as floods. And yet, communities do not always integrate their comprehensive plans with their Hazard Mitigation Plans. Comprehensive plans are often silent on the topics of hazard planning and resilience, and many Hazard Mitigation Plans do not discuss land use tools that could guide future development away from known flood hazard areas. In many communities, local planning and zoning staff are not involved in the preparation of Hazard Mitigation Plans, just as emergency management personnel are often not involved in the comprehensive land use planning process. If comprehensive plans and Hazard Mitigation Plans are not coordinated, they might inadvertently act at cross-purposes. For example, a comprehensive plan might identify future growth areas in unsafe locations if it does not take into account future flood hazard areas. Similarly, a Hazard Mitigation Plan that is not coordinated with the comprehensive plan might inadvertently recommend hazard mitigation activities in areas that are inconsistent with the comprehensive plan priorities.

To improve flood resilience, communities could better coordinate the process of developing and implementing their comprehensive plans and Hazard Mitigation Plans. They could ensure that stakeholders involved in resilience planning, such as emergency managers, also help develop the comprehensive plan and that planners help develop the Hazard Mitigation Plan. Ensuring that comprehensive plans and Hazard Mitigation Plans are integrated and consistent with each other can help decision-makers understand what infrastructure in their communities is at risk and help them outline a strategy for fostering growth in safer locations. FEMA's *Integrating Hazard Mitigation Into Local Planning* provides information about how to integrate hazard mitigation activities into local planning efforts,[11] and FEMA Region 10's *Integrating the Local Natural Hazard Mitigation Plan Into a*

Community's Comprehensive Plan: A Guidebook for Local Governments provides case studies of communities that have integrated their plans.[12] Several communities across the country have successfully integrated hazard planning elements into their comprehensive plans, including Bourne, Massachusetts and Roseville, California.[13] Some states, including Vermont[14] and Rhode Island,[15] now require communities to address natural hazards in their comprehensive plans.

Coordinating these plans and implementing the appropriate policies, regulations, and strategies to make these plans a reality can also place communities in a better position to request post-disaster assistance if and when the next disaster occurs. Communities that identify potential hazard mitigation projects and begin completing hazard mitigation grant applications before a disaster occurs, instead of having to quickly develop such lists of projects in the aftermath of a disaster, are better positioned to apply for federal funding for disaster recovery and can speed up their recovery process.

To make comprehensive plans and Hazard Mitigation Plans more effective, communities can also ensure that their capital improvement plans and budgets match the priorities outlined in their comprehensive plans and Hazard Mitigation Plans and can prioritize capital improvements that are located in safer, less vulnerable locations. This approach might mean that a community might prioritize fixing or expanding facilities and infrastructure in safer locations, or a community might choose to strengthen or relocate existing facilities and infrastructure that are located in vulnerable locations. Using these approaches can help make better use of scarce capital improvement funds while also enhancing flood resilience.

B. Conduct thorough policy and regulatory audits.

Communities might also wish to undertake a thorough assessment or audit of their zoning, subdivision, stormwater management, and other regulations. This assessment can tell the community whether current policies and regulations will let it achieve the goals in its plans, identify which policies might need to be updated, and determine where new policies could be helpful. The checklist in Appendix C can provide a starting point for communities that are interested in conducting a policy and regulatory audit to enhance resilience. Other scorecards and checklists, such as the Vermont Natural Resources Council's *Resilient Communities Scorecard*, may also help communities in Vermont and other states assess their resilience in key areas including transportation, energy, housing, land use, and healthy community design.[16] The *Smart Growth Implementation Toolkit* provides another set of tools to help communities implement smart growth approaches to development.[17] Communities might choose to review several smart growth and resilience audit tools to determine which audit (or audits) will fit their needs.

C. Amend zoning, subdivision, and stormwater policies and regulations to match plans.

After assessing existing policies and regulations, communities are well-positioned to update and amend those policies and regulations to match the goals outlined in their plans. Communities can consider several policy and regulatory options to achieve their resilience goals and can choose those options that fit their community context and tailor those policies and regulations to fit their needs. These policy options are discussed in greater detail in Section 3.

D. Consider participating in the National Flood Insurance Community Rating System.

Communities that are beginning to implement strategies to enhance their flood resilience might wish to participate in the National Flood Insurance Program Community Rating System.[18] The Community Rating System is a voluntary program that recognizes and encourages community flood plain management activities that exceed the minimum standards of the National Flood Insurance Program. One of the benefits of the Community Rating System is that flood insurance premium rates for policyholders in participating communities are discounted. The Community Rating System uses a class rating system that is similar to fire insurance rating to determine flood insurance premium reductions. Most communities enter with a Class 9 rating, which entitles policy-holders in participating communities to a 5 percent discount on their flood insurance premiums. The maximum discount is 45 percent for Class 1 communities.[19] Currently, only three Vermont communities participate in the Community Rating System, all at a Class 9 level.[20] The low level of participation might be due in part to the high administrative cost of participating, which can be burdensome for towns with few permanent staff. To decrease the administrative burden to participate in the Community Rating System, a regional organization might assist several of its communities to develop their applications simultaneously, thereby achieving an economy of scale.

3. Local Land Use Policy Options and Strategies to Improve Flood Resilience

There are several policy options that communities can consider implementing to increase flood resilience. Communities can choose which options fit their community context and tailor the policies to fit their needs. The policy options are organized into four categories (see Figure 5):

A. River Corridors: Conserve land and discourage development in particularly vulnerable areas along river corridors such as flood plains and wetlands.

B. Vulnerable Settlements: Where development already exists in vulnerable areas, protect people, buildings, and facilities to reduce future flooding risk.

C. Safer Areas: Plan for and encourage new development in areas that are less vulnerable to future floods.

D. The Whole Watershed: Implement enhanced stormwater management techniques to slow, spread, and infiltrate floodwater.

These four place types—River Corridors, Vulnerable Settlements, Safer Areas, and the Whole Watershed—describe different geographic areas within a river valley. The types of policy options and strategies that would be most effective at enhancing flood resilience will differ from place to place. For example, in river corridors, communities might focus on conserving undeveloped land to allow room for flood water to periodically inundate, while in safer areas, they might target future growth.

The policy options under these four categories offer multiple and interrelated benefits. For example, directing development out of flood plains not only keeps people and property safe, it also protects the ability of flood plains to hold and slow down flood water before it reaches downstream settlements. Ultimately, it is up to the state and communities to select the appropriate policies, adjust them to meet their specific context, and allocate resources accordingly. Each jurisdiction can weigh their resilience goals with other community priorities and can determine the best policies and approaches that will help them meet their objectives.

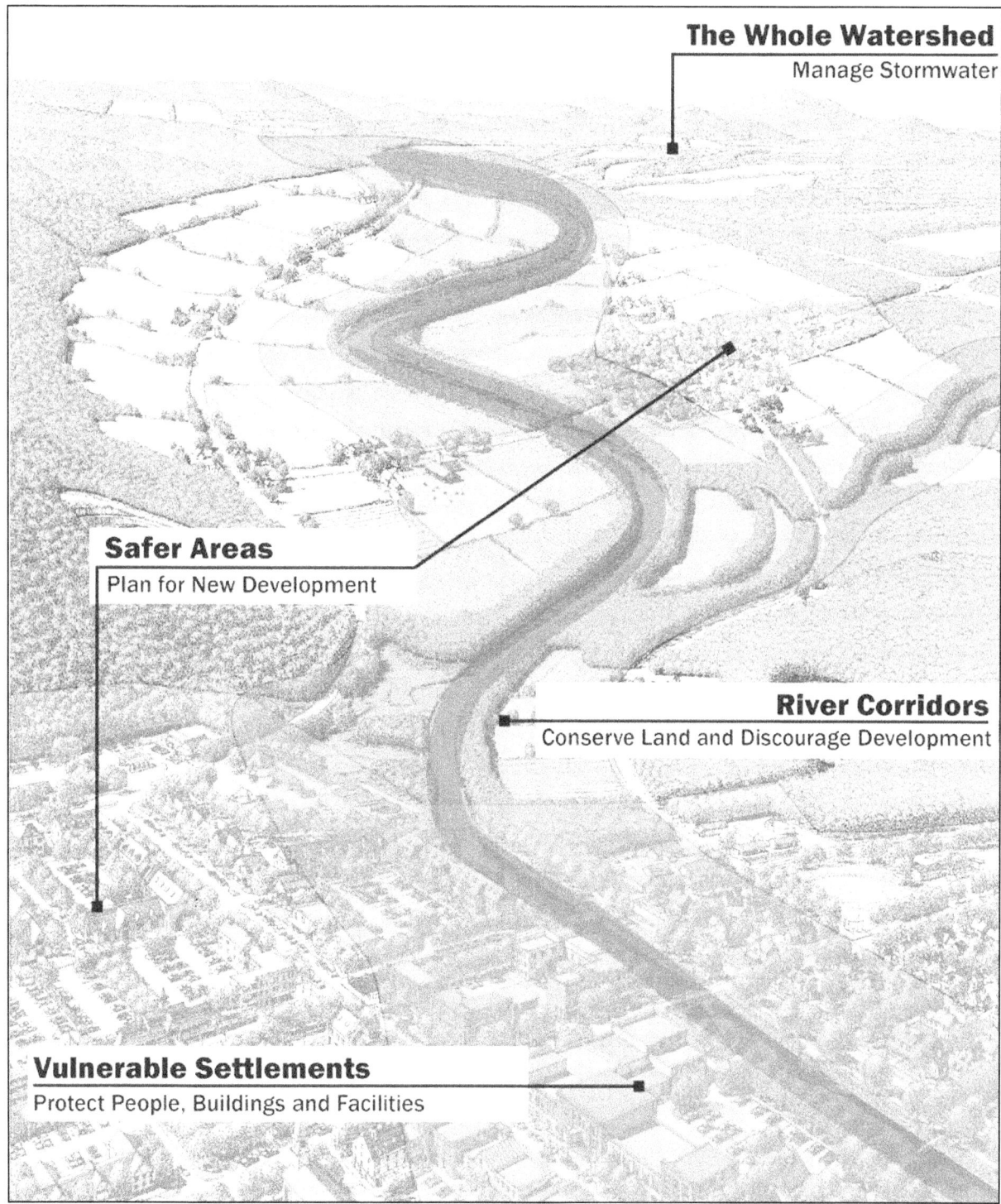

Figure 5. This graphic illustrates the four categories of approaches to enhance resilience to future floods. Credit: Vermont Agency of Commerce and Community Development.

A. River Corridors: Conserve land and discourage development in particularly vulnerable areas along river corridors such as flood plains and wetlands.

Communities that wish to reduce future flood risk can consider conserving land and discouraging development in particularly vulnerable areas, such as flood plains along river corridors. Conserving land in river corridors, especially land that is in a natural, vegetated state, can reduce flood risk by absorbing and making room for water during floods. Moreover, discouraging development in these areas can reduce the risk that homes, businesses, and critical infrastructure will be damaged by floods.

River Corridors

This section of the report corresponds with the "Conserve Land and Discourage Development in River Corridors" section in the Flood Resilience Checklist in Appendix C. Please see the checklist for a list of strategies to consider to conserve land and discourage development in river corridors.

1. Acquire or protect land in flood-prone locations.

To accommodate flood water and reduce the risk that homes and businesses will be damaged, communities can acquire or protect land in flood-prone locations. EPA's 2012 publication, *Essential Smart Growth Fixes for Rural Planning, Zoning, and Development Codes,* also provides helpful information about protecting agricultural and sensitive natural areas.[21]

Vulnerable land in river corridors can be protected in several ways.

- Purchase land or acquire conservation easements from willing sellers.
- Coordinate buyouts of properties that are repeatedly flooded.
- Develop a Transfer of Development Rights (TDR) program.
- Provide tax incentives for conserving vulnerable land.
- Restore riparian and wetland vegetation.

Communities can partner with willing landowners and land trusts or other organizations to purchase land outright or

Figure 6. Conserving land in undeveloped river corridors like this one in the Mad River Valley can help minimize risk to structures during floods. Credit: EPA.

acquire conservation easements[iii] on undeveloped properties along a river, such as a farm or forestland, to ensure that the land remains undeveloped and retains its ability to accommodate flood water. To create an acquisition program, a community would establish clear goals for the program, identify priority lands to protect based on community goals and flooding risk, and identify potential funding mechanisms. Communities that already have an acquisition program in place might need to change the program to ensure that it includes areas within the community that are vulnerable to flooding. Funding sources for acquisition programs (depending on state-enabling legislation) could include sales taxes (many communities across the United States, for example in

---

[iii] A conservation easement is "a legal agreement between a landowner and a land trust or government agency that permanently limits uses of the land to protect its conservation values." Land Trust Alliance. "Conservation Easements." http://www.landtrustalliance.org/conservation/landowners/conservation-easements. Accessed Apr. 9, 2014.

Colorado,[22] use this source), general obligation and revenue bonds, real estate transfer taxes, impact fees, and special district fees.

River corridor conservation projects in Vermont are often funded by combining a variety of sources. Potential sources of revenue include:

- A statewide 1 percent property transfer tax that is distributed by the Vermont Housing Conservation Board for conservation and housing projects statewide.[23]
- Local land trust conservation funds.
- Federal Fish and Wildlife conservation funds.
- State river management funds.
- Conservation funds managed by municipalities.[24,25]

Several Vermont communities, such as Brattleboro and Shelburne, have established local conservation funds through the authorities under state-enabling legislation for the purpose of protecting open space.[26,27] Some communities have increased the local property tax rate to provide a stream of revenue for these funds. For example, voters in Charlotte, Vermont, passed a levy to increase the property tax rate by 2 cents for 10 years to establish a conservation fund.[28,29] Williston, Vermont enacted recreation impact fees to acquire

Figure 7. This map shows land within the Mad River Valley flood plain that is protected by conservation easements. (Vermont Land Trust easements are shown in green and conserved flood plain areas are shown in green stripe.) Credit: Vermont Land Trust.

parks and open space.[30] The Vermont River Conservancy has several additional examples of conservation projects that combine several sources of funding.[31]

Communities could also work with FEMA or state agencies to identify properties that have been repeatedly flooded, and when funding is available, coordinate buyouts of those properties, remove structures on those properties, and allow the land to serve as a buffer for future floods.[32] FEMA's _Property Acquisition Handbook for Local Communities_ provides a helpful resource for communities considering buyouts.[33] Charles City, Iowa used FEMA's buyout program and other resources to turn frequently-flooded riverfront property into a vibrant, riverfront park that can help buffer from future floods and is an amenity for the community.[34]

Transfer of Development Rights (TDR) programs can also help protect agricultural lands and sensitive natural areas. TDR programs must be allowed under state law in order for municipalities to implement them. Vermont state law allows TDR programs,[35] and several Vermont communities, including Stowe, Vermont, have developed such programs. Under a TDR program, sensitive or vulnerable lands, such as flood plains or land in a river corridor, are zoned to restrict development and designated as a "sending area." Communities then designate "receiving areas" where they wish to see additional development. Those "receiving areas" are zoned to allow additional density. Landowners who own properties in a sending area are granted development credits for the development rights that have been reduced by the rezoning and can sell those credits to developers who wish to develop in a receiving area. TDR programs have been used successfully in many areas, including Maryland[36,37] and New Jersey,[38] to preserve open space and agricultural lands while compensating landowners for the change in development rights. TDR programs have often been implemented in faster-growing areas with significant development pressure, but rural regions or small towns might consider a TDR program implemented at a county or regional scale.

Communities or states could also consider providing tax incentives to protect important land.[iv] For example, in Virginia, the state legislature passed a Riparian Buffer Tax Credit in 2000 that grants a tax credit equal to 25 percent of the value of timber retained in a buffer up to $17,500. The buffer must be at least 35 feet wide and maintained for 15 years.[39] In Vermont, owners of farm and forest land can apply to participate in the Current Use program, the purpose of which is to allow the valuation and taxation of farm and forest land to be based on its remaining in agricultural or forest use instead of its value in the market place. This program can help keep agricultural and forest land in production, slow development on these lands, and achieve greater equity in property taxation on undeveloped land.[40]

To further enhance the ability of vulnerable land to accommodate flooding, some communities encourage riparian and wetland vegetation restoration. Restoring such vegetation can help absorb stormwater and decrease erosion. Restoring wetland and riparian vegetation is a major focus of Chesapeake Bay protection efforts such as stream restoration projects in Baltimore County, Maryland.[41] Federal programs, including the U.S. Department of Agriculture's (USDA) Conservation Reserve Enhancement Program, could help restore agricultural land along streams, as has been done in Vermont.[42] Several other USDA programs might also be helpful to communities that wish to conserve vulnerable land.[43]

2. **Encourage agricultural and other landowners to implement pre-disaster mitigation measures.**

Agricultural land in flood plains may be subject to erosion during floods, impacting farmers' ability to continue agricultural activities on their property. However, with planning and implementation of pre-disaster mitigation measures, agricultural land can be protected and can provide flood storage capacity during heavy rains, reducing flood-related damage and associated losses to both the farm and the community.

Figure 8. Agricultural land can help absorb flood water, particularly when landowners implement pre-disaster mitigation measures. Credit: Lars Gange & Mansfield Heliflight.

Localities can work with agricultural landowners to reduce the risk that farmland will be eroded by future floods and simultaneously reduce flood risk for the community by purchasing conservation easements on farmland or providing other incentives to agricultural landowners to implement pre-disaster mitigation measures that could reduce flooding risk. Disaster preparedness checklists such as *Ready Ag: Disaster and Defense Preparedness*

[iv] Specific incentives that communities can offer vary by state and by community.

for Production Agriculture, developed by Penn State College of Agricultural Sciences Cooperative Extension, may be useful in identifying general disaster preparedness techniques for agricultural landowners.[44]

Agricultural landowners might also consider implementing specific flood mitigation measures, such as storing hay bales in areas less likely to be flooded, since these bales can be carried into the river during floods, clogging culverts and bridges, which can create a dam downstream and inadvertently contribute to increased flooding along the riverbanks. Farmers and forestland managers can also install ponds or swales to capture stormwater and plant vegetation that can tolerate occasional inundation. Using such techniques can help reduce damage from flooding and can also help recharge aquifers. The Extension Disaster Education Network provides information on best practices and resources to reduce the impact of disasters, including flooding.[45]

3. Implement flood plain development limits that exceed FEMA requirements.

Many communities place restrictions on development in FEMA-identified Special Flood Hazard Areas. However, those designated areas do not always represent the extent of land that is vulnerable to flooding, such as in Vermont, where areas subject to fluvial erosion might be outside the mapped flood plain. Other communities regulate land use in the flood plains based only on the National Flood Insurance Program recommended standards, which allow new structures, fill, and other uses in the flood plain, as long as the development meets minimum protective standards (i.e., residential structures are elevated 1 foot above base flood elevation).[46,47]

The experiences of communities across the country demonstrate that simply adopting the National Flood Insurance Program minimum standards does not guarantee avoidance of flood damage and losses.[48] To avoid this problem, local governments could explore prohibiting all new development in flood plains or floodways. According to the National Flood Insurance Program definitions, a flood plain is "any land area susceptible to being inundated by flood waters from any source,"[49] and a floodway is "the channel of a river or other watercourse and the adjacent land areas that must be reserved in order to discharge the base flood without cumulatively increasing the water surface elevation more than a designated height."[50]

For example, in the wake of repeated flooding along the Harpeth River in Franklin, Tennessee, the town prohibited all new development in the flood plain. From a legal perspective, exceptions may be necessary in cases where already-subdivided lots are wholly within the flood plain and might have vested development rights. In such instances, development might be allowed but would be subject to higher elevation requirements (e.g., 2 or more feet above the base flood elevation) and additional waterproofing and safety standards. However, in areas subject to fluvial erosion (described below), simply elevating a structure might not reduce the risk of damage.

4. Implement fluvial erosion hazard zoning.

In some communities, erosion along rivers and streams caused by flooding is a more serious threat than flood inundation, especially in Vermont's hilly and mountainous terrain. Fluvial erosion is erosion caused by streams and rivers and can range from gradual bank erosion to catastrophic changes in river channel location and size during floods.[51] Development in river corridors can cause erosion and changes to the river channel (see Figure 9). Such erosion is particularly prevalent in narrow valleys or where streams have been altered and channelized. Fluvial erosion can destroy bridges, culverts, roads, and houses.

To further protect vulnerable land and avoid exacerbating downstream flooding, communities could explore fluvial erosion hazard zoning for land along rivers and streams. Such zoning, which is based on river corridors and flood hazard areas, can limit or prohibit development in fluvial erosion hazard areas. This technique is relatively new but is being implemented in Vermont and New Hampshire. Waitsfield, one of the Mad River Valley communities studied in this project, has incorporated fluvial erosion hazard regulations in its development codes. However, in many states the mapping necessary to implement such zoning is not yet available. Those jurisdictions might wish to conduct river corridor assessments and use the best available science and data upon which to base fluvial erosion hazard zoning.

If communities choose to allow limited development in fluvial erosion hazard areas, they could require compensatory flood storage to balance the loss of natural flood storage capacity caused by that development and thereby offset impacts on existing structures and public safety. However, this strategy might not reduce flooding risk as effectively as limiting development and redevelopment in these areas altogether.

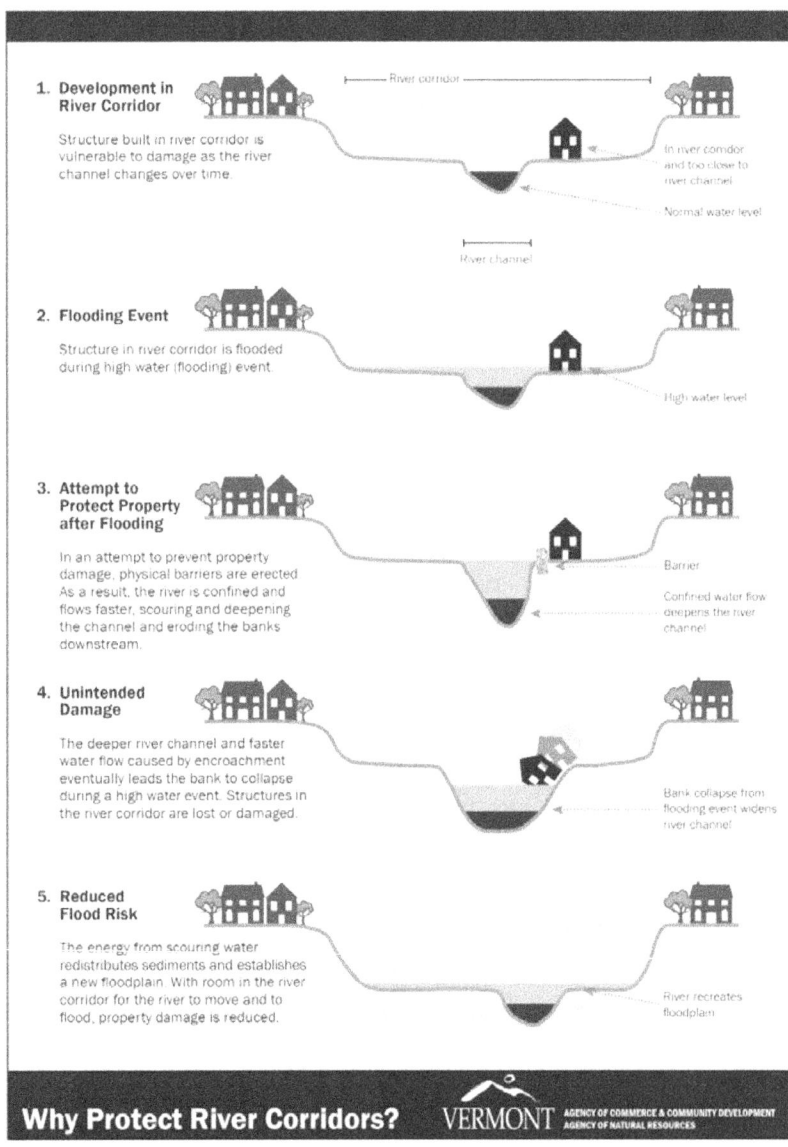

Figure 9. This graphic shows how development in the river corridor can impact the river channel. Credit: Vermont Agency of Commerce and Community Development and Vermont Agency of Natural Resources.

5. Adopt agricultural or open space zoning.

Agricultural or open space zoning is another technique available to communities that wish to protect land to allow flood water to spread and soak in the soil. This type of land use policy can limit or prohibit development in agricultural or other natural areas by limiting the number of residential units allowed on a parcel.

Figure 10: The Mad River and its surrounding landscapes make the region a beautiful place to live and visit. By planning to conserve land in the river corridor, Mad River Valley communities can also reduce damage from future floods. Credit: EPA

Some communities with agricultural or open space zoning currently allow development at densities of one unit per 2 to 5 acres. This density might inadvertently lead to spread-out, large-lot development that might fail to protect agricultural lands and open space and fail to allow effective flood storage. Increasing the agricultural or open space zoning to require a minimum lot size of 20 acres or more might more effectively preserve agricultural and open space uses and manage flood water. Many farming communities in Wisconsin and Minnesota[52,53] have adopted agricultural zoning with a minimum lot size of 20 acres or more, and Blaine County, Idaho, adopted a resource conservation zone district that allows only one unit per 160 acres.[54] Colchester, Vermont, a community near Burlington, has a minimum lot size of 25 acres in their agricultural zoning district.[55]

6. Adopt conservation or cluster subdivision ordinances.

Some communities are adopting conservation or cluster subdivision ordinances that encourage or require new development to protect tracts of intact open space (including sensitive natural areas like river and stream corridors) while clustering development into a smaller section of the parcel. Windsor,[56,57] Hartford,[58] and St. George[59] are examples of Vermont communities that have implemented such approaches. These types of ordinances might help conserve land that is important for retaining flood water. Conservation subdivisions work best when they are adjacent to existing development rather than being separated and spread out across the landscape. More information about cluster subdivision best practices can be found in EPA's 2012 publication, *Essential Smart Growth Fixes for Rural Planning, Zoning, and Development Codes*.[60]

B. Vulnerable Settlements: Where development already exists in vulnerable areas, protect people, buildings, and facilities to reduce future flooding risk.

Many historic downtowns are located along rivers and in flood plains, which often contributes to their attractive character and to the town's or region's economy. These historic downtowns represent significant investments in infrastructure over generations, and many communities choose to repair and rebuild these areas after floods because of their economic, cultural, and social importance. If communities choose to rebuild in areas that are particularly susceptible to future flooding, they can

Vulnerable Settlements

This section of the report corresponds with the "Protect People, Buildings, and Facilities in Vulnerable Settlements" section in the Flood Resilience Checklist in Appendix C. Please see the checklist for a list of strategies to consider to protect people, buildings, and facilities in vulnerable settlements.

take some steps to reduce the damage that might occur in future floods, although they cannot eliminate these risks entirely.

Changes to the National Flood Insurance Program might influence how communities consider protecting assets in vulnerable locations. In 2012, the Biggert-Waters Flood Insurance Reform Act became law. It removes subsidized rates (pre-Flood Insurance Rate Map rates) for certain classes of structures and allows rates to increase by 25 percent per year until actuarial rates are achieved. These changes will mean that premium rates will increase for some, but not all, National Flood Insurance Program policyholders. However, on March 21, 2014, President Obama signed the Homeowner Flood Insurance Affordability Act into law. This law repealed and modified certain aspects of the Biggert-Waters Flood Insurance Reform Act. FEMA is in the process of providing guidance for how the Biggert-Waters Flood Insurance Reform Act and the Homeowner Flood Insurance Affordability Act will influence flood insurance rates in the future.[61]

Figure 11. Roads, bridges, businesses, and homes in the Mad River Valley were vulnerable to flooding during Tropical Storm Irene. Credit: Clarion Associates.

1. Finance conventional protection methods.

Many communities that have experienced flooding from events like Tropical Storm Irene in Vermont or Hurricane Katrina in Mississippi and Louisiana have pursued conventional, engineered approaches to protect development in these areas, such as armoring riverbanks and coastal areas with rock riprap, channelizing rivers, and elevating structures in the flood plain. These approaches will likely continue to be used in the future but can be combined with non-structural techniques, such as planting trees and vegetation along riverbanks, to enhance their success. FEMA's publication *Engineering with Nature: Alternative Techniques to Riprap Bank Stabilization* provides examples of alternatives to engineered approaches towards streambank stabilization.[62]

One of the challenges of conventional, structural engineered approaches to flood resilience is their cost. Armoring riverbanks and rebuilding and elevating structures can be very expensive. Engineered approaches can also cause future unintended flood damage upstream and down. Riprap tends to increase the speed of water flow and can cause erosion downstream in some areas while contributing to siltation in other areas.

Due to the cost of rebuilding damaged infrastructure, communities often seek funds from the federal government for these efforts. The major federal funders include the U.S. Army Corps of

Engineers (USACE), FEMA, the U.S. Department of Housing and Urban Development (HUD), and the U.S. Department of Transportation (DOT).

The USACE builds and repairs major flood control projects such as dams and levees, sometimes requiring a state or local match for the investment. These projects can be very expensive, underscoring the need for less expensive, non-structural techniques discussed above.[63]

FEMA has several funding programs, including its Public Assistance Program, which provides local governments with funding to repair critical public infrastructure following a disaster.[64] In addition, FEMA's Pre-Disaster Mitigation Grant Program and its Hazard Mitigation Grant Program help underwrite the cost of repairing and upgrading damaged public facilities. These programs also provide funding to demolish, relocate, or elevate structures in hazard-prone areas such as Special Flood Hazard Areas.[65,66] The Hazard Mitigation Grant Program requires that projects proposed to reduce flooding risk or increase resilience be included in or compatible with their local Hazard Mitigation Plan. FEMA has other programs[67] that local governments can use to repair and upgrade their damaged public facilities.

HUD has several programs that fund infrastructure construction and repair. Many small communities have funded flood resilience-related capital improvements through the competitive Small Cities Community Development Block Grant program.[68] Local governments can, for example, use these funds for public drainage projects before a flood. After a disaster, HUD activates its Community Development Block Grant (CDBG) Disaster Recovery Funds.[69] In Vermont, HUD has delegated the administration of the HUD disaster funds to the Vermont Agency of Commerce and Community Development in accordance with its HUD-approved plan, Vermont's Community Development Block Grant Disaster Recovery Action Plan.[70] Other states might have similar state-level administration of HUD disaster funds. Flood-related projects can be implemented under the regular Community Development Block Grant program as well.[71]

Finally, both FEMA and DOT's Federal Highway Administration (FHWA) make funds available for road reconstruction due to flood damage.[72] FEMA's funds can be used to reconstruct local roads, while FHWA's funds can only be used on roads that are on the federal-aid highway system, which typically does not include most local roads.

While states do not typically have flood disaster funding programs at the same scale as the federal government, they often give some assistance to communities in the aftermath of a disaster. State agencies that fund disaster recovery and resilience usually include transportation, community and economic development, health, environment, natural resources, and agriculture agencies. See Section 4 of this report for more information on the role of states in disaster preparedness, response, and recovery.

2. **Upgrade regulations to protect vulnerable structures.**

Many communities control flood plain development through special flood plain or flood hazard area zoning overlay districts with associated development standards. Many of these standards require the lowest floor of any structure in these districts to be elevated at least 1 foot above the base flood elevation. Base flood elevation is the elevation to which flood water is expected to rise during a 100-year flood (a flood that has a 1 percent chance of being equaled or exceeded in any given year).[73] Communities could consider increasing this requirement to a minimum of 2 or more feet above the base flood elevation to provide an extra margin of safety, although as noted above, this may not be sufficient in some places such as fluvial erosion hazard zones (see Section 3.A.4). Lake County, Illinois,[74] and Fort Collins, Colorado,[75] have implemented these enhanced requirements, and the State of New Hampshire's model flood plain protection ordinance incorporates this approach.[76] While these enhanced standards might help protect structures in frequently flooded areas, these requirements alone might not eliminate flooding risk entirely, particularly since climate change projections suggest that floods will intensify in most regions of the United States, especially in the Midwest and Northeast.[77] The Land Use Institute's report, *Preparing for the Next Flood: Vermont Floodplain Management*, discusses the legality of these enhanced standards and other stormwater management regulations.[78]

Alternatively, towns could consider prohibiting development in the floodway or flood plain entirely to reduce risk further (see Section 3.A.3). Communities could also establish a temporary building moratorium on all new development after a flood occurs, allowing time to ensure that new development will be compatible with the community's goals.

3. **Address nonconforming uses.**

Regulations for nonconforming structures and uses might also affect a community's flood resilience. Many communities commonly place zoning and building code controls on the expansion or renovation of nonconforming structures and uses, with a goal of replacing or removing these structures over time. If a nonconforming structure or use that does not meet these standards is reconstructed or redeveloped following significant damage—"significant" typically means that repair costs exceed a dollar amount or percentage of the structure's value specified by the local government—the new structure or use is required to be in full compliance with all current standards, including setbacks, height, and lot area. Nonconforming use zoning rarely allows any type of expansion, including elevating a building to make it more flood resistant.

While these nonconforming use regulations make sense in many circumstances, they can have unintended consequences in areas that have been or might be subject to major storm damage. Because full compliance with current standards might be costly, property owners might choose to undertake only minor repairs to make their structures habitable rather than invest in major renovations that might trigger nonconformity provisions. This unintended consequence of nonconformity provisions might lead to less investment in a storm-damaged area and might mean that property is still vulnerable to future floods. Local governments also might have complicated approval procedures for

Figure 12. There was extensive cleanup work in Moretown, Vermont after Tropical Storm Irene. Credit: Stephen Magill, Vermont Department of Environmental Conservation.

renovations or expansions on nonconforming properties, which creates another hurdle to economic recovery in storm-damaged areas.

Many areas of the country were developed before implementation of the National Flood Insurance Program. As a result, many communities have large stocks of development that do not comply with current flood damage prevention requirements. Often these homes and businesses fail to comply with zoning-related requirements such as setbacks, off-street parking, or design-related provisions. Because modifications to these older structures would trigger the requirement for full compliance with all development standards, which can be cost-prohibitive, these nonconformities continue unchanged through the years. Standards that allow identical replacement of these nonconforming structures following storm events are politically popular but do little for the community's long-term flood resilience.

To address these problems, some communities are implementing nonconforming use regulations that recognize partial compliance with development standards and incorporate incentives for property owners to redevelop and/or reconstruct nonconforming structures using more hazard-resilient techniques, such as building elevation or flood-proofing of heating, ventilation, and air conditioning (HVAC) equipment (see Figure 13). Incentives for redeveloping nonconforming structures, when coupled with requirements for greater hazard resilience, can help development in flood-prone areas better withstand future floods.

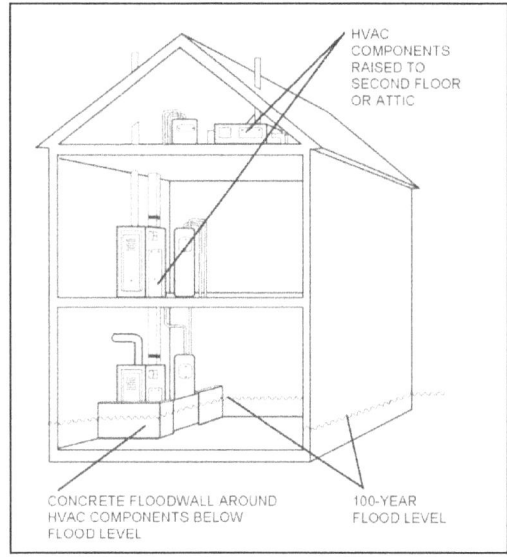

Figure 13. HVAC equipment can be raised or flood-proofed in buildings located in areas at high risk of flooding. Credit: FEMA.

Modifications to the nonconforming provisions that also provide an incentive for redevelopment (for example, expanding a building's floor area) can help home and business owners justify the costs of achieving compliance and can foster redevelopment that is more consistent with current zoning and building codes. Coupling these incentives for redevelopment with requirements for partial compliance with key development regulations (e.g., flood damage prevention standards within special flood hazard areas) can improve overall flood resilience more than if full compliance with all development regulations was required. In this situation, both the property owner and the community reap benefits. The home or business owner can increase the value of their property without incurring the expenditure of full code compliance, while the community benefits from a structure that is less likely to sustain serious damage during a future flood.

4. **Upgrade or adopt building codes to promote safer development.**

Adopting building code requirements for structures built or reconstructed in or near flood plains can help protect structures and people. The way states handle building codes varies from state to state. Some states have statewide codes and leave little opportunity for communities to adopt more stringent codes, while other states delegate building codes entirely to the local community's jurisdiction. In Vermont, the state administers building codes for commercial buildings and multi-family housing, but not for single family homes. The state also allows local jurisdictions to have stricter building codes than what the state requires and allows municipalities to adopt codes for single family homes.[79,80,81] When local jurisdictions have control over their building codes and have the resources to administer such codes effectively, they could consider upgrading their standards to provide an extra margin of safety from flood damage. The International Building Code and International Residential Code, which most state building codes adopt or use as a foundation, reference FEMA, the National Flood Insurance Program, their maps and information, and the American Society of Civil Engineers' Flood Resistant Design and Construction Standards 24-05—all of which require higher design and construction standards for flood-prone areas.[82,83]

While each state and local jurisdiction has differing laws governing local authority to adopt or modify building codes, most local governments in the United States have the legal authority to adopt zoning provisions that respond to varying levels of risk, including those related to flood and weather variability. Thus, communities can use zoning codes, including overlay zone districts, as an alternative to building codes to enact requirements for flood mitigation and flood-proofing activities.

If local governments have limited authority to vary from state-imposed building codes and do not choose to use zoning codes to enact flood mitigation requirements, they could provide incentives such as increased density or building height for the voluntary use of flood-resistant design and building standards, such as those outlined in the International Green Construction Code.[84]

5. **Create new flood storage capacity through redevelopment.**

When redevelopment opportunities arise in vulnerable areas next to rivers, communities can require developers to design projects to include additional flood storage capacity. New flood storage capacity could mean creating parks and other open spaces in flood-prone locations, replacing a vertical wall along a river bank with a more gradual slope to create more room in the river channel for rising water, creating a shallow depression in a lawn that can accommodate inundation, or designing buildings to enable the first floor or basement to flood (and then be readily repaired when the waters recede). Localities can encourage developers to create flood capacity in new development by providing density bonuses or reduced stormwater fees in exchange for creating flood capacity improvements on site or zoning overlays that indicate where new development must include additional flood capacity features.

6. **Help people connect with the river.**

In some historic, riverfront towns and villages, the development faces away from the river, including some communities in Vermont. Except at bridge crossings, residents might rarely see or consider the river as a part of community life—until a flood occurs. A river can be a social and economic asset if residents can safely access the riverfront. Opportunities to see and engage with the river could increase residents' consciousness of the river's presence and motivate them to engage in planning for future flooding and river protection.

In vulnerable settlements, communities can consider creating parks, outdoor dining and vending, river-based recreation like fishing and kayaking, and other activities that can withstand flooding and bring people closer to the river during normal flows. Implementing these approaches can also provide important economic development opportunities for communities. A 2009 publication developed by EPA, the National Oceanic and Atmospheric Administration, Rhode Island Sea Grant, and the International City/County Management Association, *Smart Growth for Coastal and Waterfront Communities*, provides a set of smart growth guidelines for waterfront development.[85] The tools, techniques, and examples described in the publication provide specific ideas for how Vermont communities could target their efforts to promote flood-friendly uses along the riverfront.

7. **Relocate structures to less vulnerable areas.**

As certain structures are flooded time and again, some communities and property owners might determine that it would be preferable to relocate them or rebuild them in safer areas. The decision to relocate can be difficult, emotional, and expensive, and it is usually a last resort for a community whose residents may be reluctant to leave their homes and move their businesses. Relocating can impose a disproportionate burden on low-income people in the community who often live or own businesses in vulnerable areas. Making a concerted effort to engage low-income, minority, and underserved community members in any discussion of relocation can help ensure their concerns are well-understood and that they are informed about their risks.

When the community decides to relocate structures through extensive and thorough community outreach, local governments can make the process easier for those who choose to relocate by creating a coordinated package of relocation services and resources for residents, including financial and logistical assistance with relocation. Through the relocation process, local governments can also move critical facilities such as town halls, fire and safety facilities, and drinking water facilities to less vulnerable locations, if possible.

Some communities have created funding mechanisms to buy properties that are susceptible to future floods. For example, in Napa County, California, the community voted to institute a ½ cent sales tax to pay the local share of a federal flood control project that includes acquiring flood-prone properties.[86] Stormwater utility fees could also be used for this purpose (see Section 3.D.1). According to a study by the University of Maryland, a stormwater utility fee of $20 per residential unit could generate from $500,000 for counties with 25,000 households to $10 million annually for counties with 500,000 households. The revenue generated from such stormwater utility fees could be used to purchase flood-prone properties, which can protect other properties within the community.[87]

Figure 14. Protecting assets in Waitsfield, Vermont, is critical to maintaining the town's resource- and recreation-based economy and quality of life. Credit: Lars Gange & Mansfield Heliflight.

Local jurisdictions could choose to create a pre-disaster anticipatory relocation fund when such a program is cost-effective (i.e., if the costs of anticipatory relocation are presumed to be less than the costs of post-disaster relocation). Communities considering this approach could first prepare a relocation assessment to identify:

- The range of uses, services, and facilities eligible for funding;
- Priorities for protecting vulnerable areas;
- Potential impacts of the anticipated event (e.g., flooding) to both people and structures; and
- The potential total funding created by each available source.

Communities could also prepare a cost-benefit analysis for structures and infrastructure that compares:

- Anticipatory relocation.
- Post-impact relocation.
- Status quo with no further action needed for the damaged service or infrastructure.

If a local government is considering offering relocation as an option for residents, it must ensure that all community members are informed of their risks and involved in determining their options for relocation, should they wish to relocate. Based on the outreach results, the local government can determine the funding it will need to meet the community's wishes regarding relocation.

For a relocation fund to be successful, it must be backed by a long-term, reliable funding source such as a dedicated sales tax. Additional available funding sources might include annual appropriations from the general fund, bond issuance, and potentially, where funding will be used to relocate infrastructure, a tap fee or stormwater utility fee. If vulnerable areas will be converted to natural or open spaces, funding might also be available from foundations, nonprofit land preservation organizations, federal government grants, or local or regional parks and recreation budgets. In addition to local funding, there might be opportunities to leverage federal assistance, such as FEMA's Pre-Disaster Mitigation Grant program (authorized by the Stafford Act), Hazard Mitigation Grant Program, or Flood Mitigation Assistance Program; and HUD's CDBG program. The funds collected as part of the relocation fund can be provided directly to recipients as grants or could be used to underwrite low-interest loans for relocation costs.

C. **Safer Areas:** Plan for and encourage new development in areas that are less vulnerable to future floods.

Communities seeking to enhance their resilience to future floods can identify areas that are less vulnerable to flooding, where growth can occur more safely in the future. By encouraging development in these safer growth areas, communities can accommodate new growth while reducing flooding risk. After communities have identified *where* they can more safely grow in the future, they can then also shape *how* development is built in those locations by using the smart growth

Safer Areas
This section of the report corresponds with the "Plan for and Encourage New Development in Safer Areas" section in the Flood Resilience Checklist in Appendix C. Please see the checklist for a list of strategies to consider to plan for and encourage new development in safer areas.

principles. Several approaches and policies can help communities' direct growth into safer locations.

1. **Identify locations suitable for development and redevelopment that are safer from flooding.**

   Many communities have identified locations where future growth is desired for a variety of reasons, such as having access to existing infrastructure and/or being contiguous to other development in the community. However, some of these desired future growth areas may not be in safe locations. Communities that are interested in targeting growth in *safer* locations would need to ensure that their desired growth areas are also located in areas that can more safely accommodate growth. They can then identify these safer growth areas in the land use plan or comprehensive plan. Bringing residents, property owners, and other stakeholders together to develop a vision for how the community might accommodate new development in these locations can be very helpful. The community can incorporate that vision for future development into the comprehensive plan, revise existing regulations or adopt new regulations necessary to implement the plan, and plan new public facilities with the vision in mind.

   To identify where growth can occur more safely in the future, communities will need information about where flooding has occurred in the past and, to the extent possible, projections for future flooding that take climate change into account. The Vermont Agency of Natural Resources is developing a comprehensive website for municipalities with mapping resources that will be completed later in 2014 and could be used for this purpose.[88],[89] Designating new nodes for development, including the desired density and mix of uses for those new nodes of development, in the community's land use plan shows developers which locations the community has identified as a priority for expansion. If developers understand where the community wants to grow, they may be more likely to propose development in those locations.

2. Steer public policy and investments to support development in safer locations.

Once communities have identified locations that are safer for development, they can adopt and implement policies and make public investments that will encourage development in those safer locations.

Localities can update their zoning and subdivision regulations to remove barriers to development in safer areas. If the local plan calls for more compact development in safer growth areas, local governments can ensure that land use regulations do not unintentionally inhibit development there. For example, if regulations do not allow multifamily developments or restrict the size or height of multifamily buildings, they might make it difficult to construct medium-density developments that might be appropriate for the area. Similarly, larger front setback requirements and off-street parking standards require more land and can increase the cost of development. Revising these requirements can make streets more attractive and safer for pedestrians and bicyclists.[v]

Smart Growth Principles

Based on the experience of communities around the nation, the Smart Growth Network developed a set of 10 basic principles:

- Mix land uses.
- Take advantage of compact building design.
- Create a range of housing opportunities and choices.
- Create walkable neighborhoods.
- Foster distinctive, attractive communities with a strong sense of place.
- Preserve open space, farmland, natural beauty, and critical environmental areas.
- Strengthen and direct development towards existing communities.
- Provide a variety of transportation choices.
- Make development decisions predictable, fair, and cost effective.
- Encourage community and stakeholder collaboration in development decisions.

Source: Smart Growth Network. "Why Smart Growth?" www.smartgrowth.org/why.php

Localities can also direct public investments in new infrastructure, facilities, and schools into safer locations, which might help attract additional private investment in these areas. They can also coordinate local capital improvement plans with community plans, ensuring that maintenance and repair of existing infrastructure, as well as future capital improvements such as roads and utilities, are located in safer areas. By prioritizing capital investments such as sewer, water, and streetscape improvements in safer areas, communities can provide incentives for development to locate there.

Furthermore, communities can apply the smart growth principles (see text box) to ensure that all new development that is built in safer locations is also compact, walkable, and has a range of transportation and housing opportunities for residents. Using the smart growth principles can help ensure that future growth is both safe *and* smart.

D. The Whole Watershed: Implement enhanced stormwater management techniques to slow, spread, and infiltrate flood water.

Communities can also implement policies to more effectively manage stormwater throughout the entire watershed. Adopting these policies can help slow stormwater, spread it out over a larger area, and allow it to infiltrate into the ground rather than running off into nearby streams and rivers.

The Whole Watershed

This section of the report corresponds with the "Implement Stormwater Management Techniques throughout the Whole Watershed" section in the Flood Resilience Checklist in Appendix C. Please see the checklist for a list of strategies to consider in implementing stormwater management techniques to slow, spread, and infiltrate flood water.

---

[v] For example, many communities require an off-street parking space for every 200 or 300 square feet of commercial building when one per 400 square feet will meet parking demand, especially in smaller jurisdictions.

1. **Explore watershed-wide stormwater management.**

   Flood damage mitigation measures, such as constructing levees or armoring banks, that are implemented in one jurisdiction in a watershed can have unintended consequences for other communities in that watershed by speeding the flow of floodwaters downstream. Recognizing this, some communities, including those in Chittenden County, Vermont, are joining together to take a regional, watershed-wide approach to stormwater management.[90] To do this, communities can develop educational programs and stormwater master plans for their watersheds and use hydrologic data and watershed modeling to understand more clearly what actions to take to absorb and slow down stormwater across the watershed to reduce flooding risk.[91]

   Some communities create stormwater utilities to address stormwater management across a wider geographic area. A stormwater utility is an entity established to generate and administer a dedicated source of funding for stormwater pollution prevention activities. Generally, users pay a fee to the utility based on land use and their contribution of runoff to the stormwater system.[92] Stormwater utilities can oversee stormwater management regulation and can help prioritize, coordinate, and finance critical pre-disaster mitigation efforts such as streambank restoration projects. A 2009 EPA publication, _Funding Stormwater Programs_, provides information on ways that communities can finance stormwater management programs, the steps involved in establishing a stormwater utility, and the advantages and disadvantages of different types of stormwater utilities.[93] The report includes case studies from South Burlington, Vermont, and Newton, Massachusetts.

2. **Better manage stormwater from roads, driveways, and parking lots.**

   Roads, driveways, and parking lots made of impervious surfaces do not allow stormwater to infiltrate back into the ground and can increase stormwater runoff volumes, especially during heavy rains. In addition, the runoff collects the debris, oils, and pollutants from these paved surfaces and carries them into surface waters. Communities could consider implementing policies that can reduce the effect that roads, driveways, and parking lots have on exacerbating flooding and degrading water quality. They could encourage the use of pervious material in new driveways and parking lots, and in new roads where feasible. Geo-synthetic materials that are pervious and washout resistant can also be used for roads and can be funded using FEMA's Public Assistance Program. In addition to green infrastructure practices such as pervious pavement and roadside swales that allow stormwater to infiltrate into the ground, communities could also require that culverts, which are often too small to adequately drain stormwater from large storms, be upgraded to protect roads from damage during flooding. The state of Vermont offers guidance to towns and cities about adequate culvert sizing.[94] Where possible, communities can consider using open-bottom stream overpasses instead of culverts, since culverts can be damaging to passage of fish and other aquatic organisms. Vermont law, for example, prohibits the creation of obstructions in streams that prevent the passage of fish unless authorized by permit,[95] and the state's design standards for road crossings over streams now generally results in an open-bottom or box culvert design that allows for such passage.[96]

   In many rural communities, roads and parking areas are made of gravel, rather than asphalt. Communities often surround gravel roads and parking areas with ditches that drain and protect the surfaces during heavy rains, but these ditches might also increase flooding by conveying stormwater directly into streams and rivers. Communities can require techniques to slow the flow of water by spreading it into vegetated areas and infiltrating it in areas with pervious soils. Communities can also provide information about stormwater management techniques that private landowners could use for their driveways. Such techniques not only reduce flooding risk but can also improve water quality.

3. **Adopt local stormwater management regulations that allow the use of green infrastructure techniques.**

    While some communities in the United States have implemented comprehensive stormwater management regulations to comply with EPA or state requirements, other smaller, rural jurisdictions might not be required to implement such regulations. In Vermont, state stormwater permits are required only for developments with more than 1 acre of impervious surface and sites that disturb more than 1 acre through the stormwater program. Additionally, comprehensive state land use regulations require a review of stormwater impacts for subdivisions of 10 or more lots, commercial development of 10 or more acres, and any development at elevations above 2,500 feet. However, stormwater runoff from developments with, for example, less than 1 acre of impervious surface on a steep slope might also contribute to flooding problems. Recognizing this, some localities and regions are going above and beyond federal or state stormwater requirements to regulate stormwater throughout their communities. Williston,[97,98] Rutland,[99] and South Burlington[100] are three Vermont communities that regulate stormwater on smaller parcels or in vulnerable areas such as steep slopes or land near lakes and streams.

    Communities that want to improve stormwater management can consider requiring new developments to prepare stormwater management plans that use best management practices suggested by federal, state, or other agencies. "Hard" engineering solutions such as underground cisterns are often used to meet these requirements, but "soft" green infrastructure approaches such as ponds, swales, or wetlands could be considered as an alternative or supplement to structural solutions. Green infrastructure is an approach that uses vegetation and soil to manage rainwater where it falls (see Figure 15). It can help retain and/or reuse stormwater near where it is generated and can be less costly and less environmentally damaging than conventional stormwater treatment, particularly when it is designed into development from the start.[101] Specific green infrastructure approaches include:[102]

    - Reducing the amount of impervious surface by designing parking lots and other paved surfaces so that they are smaller.
    - Reducing the effect of impervious surfaces by directing the runoff into features where it can infiltrate, such as rain gardens; depressed landscape islands in parking lots (instead of mounded landscape islands); or bioswales, which can be located in public right of ways.
    - Using pervious concrete, pavement, or pavers in appropriate locations such as some parking lots and driveways. To maintain the perviousness of the materials, however, it is important that these surfaces be maintained by practices such as vacuum sweeping.
    - Reusing rainwater for landscaping, gardening, or irrigation (i.e., rainwater harvesting) on industrial, institutional, commercial, or residential lots.
    - Promoting the use of rain barrels to capture rainwater for later use.
    - Constructing green roofs (vegetated roofs that absorb stormwater) or blue roofs (non-vegetated roofs that are designed to store water)[103] where a controlled flow system allows water to collect and then gradually drain away.
    - Including storage underneath parking lots, streets, and sidewalks that empties through small holes into the sewer system or infiltrates into the ground.

Figure 15. Green infrastructure techniques such as rain gardens (left) and rain barrels (right) retain stormwater runoff on site and reduce peak flooding. Credit: Clarion Associates.

The town of Williston, Vermont, has adopted stormwater management regulations that incorporate many of these concepts in the "Watershed Health" section of its Unified Development bylaw.[104] The regulations include standards for holding pre-construction meetings with town staff to discuss erosion control, avoiding development on steep slopes, and requiring buffers around wetlands and along streams.

4. Adopt tree protection measures.

Large trees can absorb significant amounts of rain and can reduce stormwater velocity. To protect trees, communities could start by preserving existing, undeveloped forested areas. Communities could also require that larger trees, such as those that are more than 8 inches in diameter, be preserved on a development site as much as possible. Or, if those trees must be removed, a community could require that they be replaced at a minimum one-to-one basis[vi] on site or mitigated through payment into a municipal tree protection fund.[105] Communities could also implement requirements to retain a specified percentage of the tree canopy on a development site. For example, for a parcel that has 100 percent tree canopy cover, regulations might be designed to require that development on the site be placed so that 75 percent of the canopy is preserved. Currituck County, North Carolina, and Folly Beach, South Carolina, have tree protection codes that illustrate these approaches.[106,107] The town of Wellesley, Massachusetts also has a tree protection ordinance that may serve as a useful model.[108] Additional standards can protect trees during construction, such as requiring fencing at the tree dripline, which is the area bounded by the outer circumference of the tree branches, and where most of the roots are located.

5. Adopt steep slope development regulations.

Development on steep slopes can cause erosion and can increase stormwater volumes (see Figure 16). However, regulation of steep slope development varies widely in communities across the nation. Some communities with a history of landslides, mudslides, or earthquakes have implemented standards that prohibit building on steep slopes or reduce the density of residential development allowed in those areas. But many other communities merely caution against building on steep slopes or are silent on the topic. Some communities are beginning to recognize that

---

[vi] For example, if a tree that measures 6 caliper inches is removed, it must be replaced with a total of 6 caliper inches of new trees.

development on steep slopes can affect stormwater volume and erosion and are adopting standards that discourage or prohibit development on very steep slopes (steeper than a 30 percent grade). Williston, Vermont's steep slope development regulations reduce allowable densities on slopes greater than 15 percent and prohibit development on slopes greater than 30 percent in most instances.[109] Salt Lake County, Utah, has adopted similar regulations in a more urban context, but the regulations provide more flexibility to accommodate infill development.[110] Before adopting steep slope development regulations, localities should reach out to engage affected property owners during the process of developing the regulations, since the regulations might reduce the developable portion of certain properties. As discussed in Sections 3.A.1 and 3.A.6 of this report, non-regulatory approaches like transfer of development rights programs and conservation subdivisions can complement regulatory approaches and can also help address landowners' concerns about reducing the developable portion of their lands.

Figure 16. Development on steep slopes and poor erosion control methods can cause erosion and increase the quantity of stormwater runoff. Steep slope development regulations can help prevent some of these impacts. Credit: Vermont Stormwater Program.

6. **Adopt riparian and wetland buffer requirements.**

Stream and wetland buffer standards require development setbacks from rivers and other water bodies. These buffers can allow stormwater to infiltrate into the soil, reducing flood flows downstream in more developed areas of the community and reducing erosion by stabilizing river banks. Buffers can also remove some pollutants that would otherwise run off into local rivers. Studies show that in more rural areas, a buffer of 100 feet can significantly reduce stormwater runoff and improve water quality.[vii,111] Smaller buffers of 25 to 50 feet might be appropriate in more developed areas if supplemented with enhanced stormwater management techniques such as additional vegetation or underground cisterns. Effective buffer regulations typically include vegetation requirements in riparian areas along streams and rivers. If vegetation is stripped out of buffers, stormwater is less likely to spread and infiltrate in, and erosion will be greater. Consequently, many local ordinances add standards to protect vegetation, such as requiring construction fencing around buffer areas and larger trees and prohibiting storage of construction materials in buffer areas, which compacts soils and can damage trees and lead to additional runoff. Watershed buffer standards adopted by the town of Williston, Vermont, require a 150-foot development setback from most lakes and streams.

[vii] An accepted rule of thumb is that a stream buffer should be a minimum of 50 feet wide and preferably 100 feet to remove sediment, increase stormwater infiltration, and protect wildlife.

4. State Policy Options and Strategies to Improve Flood Resilience

While a community's flood resilience is influenced strongly by local land use decisions and local disaster resilience planning efforts, state-level policies also affect local flood resilience. This section describes ways that state policies influence flood resilience and outlines options that state agencies can consider as they seek to improve local communities' abilities to enhance their flood resilience.

Recognizing that state policies might influence communities' abilities to implement flood resilience practices, this project assessed state policies in Vermont and offered options for how those policies might be amended or new policies created to enhance flood resilience at the local level.

States influence flood resilience in a variety of ways:

- They are often responsible for coordinating disaster preparedness, response, and recovery efforts.
- They often help communities develop the capacity to prepare for, respond to, and recover from disasters.
- They serve as a conduit for resources and technical assistance from federal agencies like FEMA that provide disaster-related planning and recovery assistance.
- They implement policies that shape the universe of how local land use decisions are made and, thereby, indirectly influence communities' flood resilience.

Figure 17. States can support local flood recovery and long-term flood resilience through agency policies and coordination. Credit: Richard Amore, State of Vermont.

- Finally, they make decisions about the location and type of many infrastructure investments in communities through grants or direct provision of transportation, housing, and water and wastewater infrastructure, all of which might affect local communities' flood resilience.

The policy options that follow are offered as a starting point from which Vermont state agencies can begin to determine how they should proceed. Ultimately, it is up to the state to select the appropriate policies from this list, refine them, and allocate resources accordingly. The options fall into two major categories: A) those actions that several agencies can take together, and B) those actions that are specific to individual agencies. These policy options were developed in consultation with Vermont state agencies through the course of this project. A more detailed description of these options and action steps is available in a state policy options report, *Vermont State Agency Policy Options: Smart Growth Implementation Assistance Program, Disaster Recovery and Long-Term Resilience Planning in Vermont*, that was prepared as part of this project and is available online.[112] The state of Vermont is now setting priorities for action based on these policy options, some of which are summarized below.

A. Inter-Agency Policy Options to Enhance Flood Resilience

When state agencies are considering ways to enhance communities' flood resilience, they might wish to evaluate the prospects of potential policies in terms of their technical, administrative, fiscal, and political feasibility. This analysis could also consider current state capacity and conditions, the potential impact of the policies on flood resilience, the duration of the action (how long it will take to develop and implement the approach), and other considerations such as whether implementing the policy will create the opportunity to leverage other resources and existing policies.[113,114]

1. **Conduct an audit of state programs.**

   State agencies (including those listed in Section 4.B, below) could conduct an audit of their programs to assess the degree to which they directly or indirectly help or hinder the state's ability to improve communities' resilience. One resource to assist with such an audit is the *State Disaster Recovery Planning Guide*, developed by the University of North Carolina at Chapel Hill's Coastal Hazards Center of Excellence to help states develop, maintain, and implement state disaster recovery plans.[115]

2. **Develop a comprehensive pre-event recovery plan.**

   State agencies could also develop a comprehensive pre-event recovery plan in advance of the next disaster and set up periodic exercises to practice implementing the recovery plan. According to the *State Disaster Recovery Planning Guide,* one of the best reasons to develop a pre-disaster recovery plan is to prepare state agencies and others to act quickly after a disaster and to minimize the damage that the disaster might cause. Pre-disaster recovery plans typically address issues such as interorganizational coordination, communication, staffing, and capacity building, as well as topics such as debris management and reconstruction, so that the state is prepared to recover quickly should a disaster occur. State agencies could reach out to Community Planning and Capacity Building personnel in FEMA's regional offices[116] to develop approaches to enhance state and local partners' abilities to plan for, manage, and implement disaster recovery activities.

3. **Develop a post-disaster personnel plan.**

   State agencies can also develop a post-disaster personnel plan that describes anticipated personnel needs should a disaster occur and identifies the resources that can be provided by a network of partners, including federal and local officials, nonprofits, quasi-governmental organizations, consulting firms, and other groups. These personnel plans could emphasize the pre-disaster development of a group of trained personnel that can be prepared to assist state recovery activities should a disaster occur. While few such personnel plans exist at the state level, several counties have these plans. For example, Hillsborough County, Florida's personnel plan includes 17 county agencies and major nongovernmental organizations such as the Red Cross.[117]

4. **Map flood plains and adopt a No Adverse Impact standard.**

   State agencies can implement a comprehensive state corridor and flood plain protection program guided by the principle of "No Adverse Impact." According to the Association of State Flood Plain Managers, "No Adverse Impact flood plain management takes place when the actions of one property owner are not allowed to adversely affect the rights of other property owners…in terms of increased flood peaks, increased flood stages, higher flood velocities, increased erosion and sedimentation, or other impacts the community considers important."[118] The No Adverse Impact approach is a framework of techniques and tools that communities can use to identify hazards in their communities and identify ways to reduce those hazards, through hazard identification, planning, infrastructure, emergency services, regulations and standards, corrective actions, and education and outreach.[119] As a first step, agencies could develop and maintain a statewide river corridor and flood plain mapping program supported by flood and fluvial erosion hazard risk assessments. With statewide maps in place, state agencies could integrate a development standard of No Adverse Impact into their policies and programs. The state also could encourage municipalities to adopt No Adverse Impact standards by providing model language that could be incorporated into local regulations limiting development in flood-prone areas.

B. Agency-Specific Policy Options to Enhance Flood Resilience

In addition to coordinating efforts across multiple agencies, individual state agencies could take steps to enhance flood resilience at the local level. This section includes policy options for:

1. Natural resources and environmental protection agencies.

2. Transportation agencies.

3. Emergency management agencies.

4. Commerce, community, economic development, and housing agencies.

5. Agriculture agencies.

6. Disaster recovery offices.

1. Natural resources and environmental protection agencies could:

- Ensure that river corridor, inundation, and flood plain data is available for communities and that the process used to make decisions about river corridor and fluvial erosion hazard areas actively engages local partners that have a deep, locally grounded understanding of flood hazard risk, including how risks may change in light of climate change projections. Charlotte-Mecklenburg County, North Carolina, used an inclusive, process-oriented approach to assessing flood hazard risk that could be emulated at the state level.[120]

- Encourage or require communities to regulate their flood plains based on Flood Insurance Rate Maps and river corridor maps through a combination of setbacks, fluvial erosion hazard overlays, river corridor protection plans, best management practices, land use and Hazard Mitigation Plans, infrastructure management initiatives, and stormwater management plans. One community that has already taken action is Bennington, Vermont, which adopted a Fluvial Erosion Hazard Overlay District that requires a zoning permit that limits uses in the overlay district, prohibits certain hazardous uses, and exempts agricultural activities. This overlay district could be used by the state as a model ordinance for other communities.[121]

- Take the lead in establishing state minimum "No Adverse Impact" standards that municipalities would be encouraged or required to incorporate into local bylaws limiting development in flood-prone areas.[122]

2. Transportation agencies could:

- Incorporate hazard mitigation and flood resilience practices into project design and prioritization procedures.[123] For example, transportation agencies could ensure their designs account for flood hazard vulnerability and the effects of designs on downstream flooding and fluvial erosion, and incorporate those parameters into documents such as the Vermont State Design Standards. The vulnerability criteria used to shape resilient design parameters could be developed in coordination with natural resource agencies and regional planning organizations.

- Review all infrastructure programs, including grant programs for communities, to look for opportunities to create local incentives and prioritize projects and maintenance strategies that reduce the risk of future flood damage in vulnerable areas. Infrastructure resilience features include redundant systems; robustness (inherent strength/resistance); resourcefulness (capacity to mobilize needed resources); and rapidity (speed with which disruptions can be overcome and services restored). An example of a local grant program that can provide incentives for change is Vermont's Flood Resilience Community Program.

- Conduct and maintain an inventory system of federal, state, and local culverts. Once the inventory is complete, the results could be incorporated into the state Hazard Mitigation Plan and linked to the state's strategy for reducing risks from inadequate culverts. Vermont's transportation agency has initiated a state-wide inventory of culverts on state roads. The next step will be to coordinate with towns and regional planning organizations to evaluate town-owned structures. New York and Ohio have manuals for inspecting and inventorying state culverts that could be models for other states.[124,125]

- Coordinate with environmental and emergency management agencies and local officials to identify appropriate hazard mitigation measures, including those that might be eligible under FEMA's Public Assistance 406 Program and the Hazard Mitigation Grant Program. Measures might include increasing the size of inadequately sized culverts that were damaged during extreme events, limiting upstream development, creating catchment areas, and conducting flood engineering studies that could inform which hazard mitigation measures are appropriate.

3. **Emergency management agencies could:**

- Host statewide hazard mitigation workshops emphasizing the link between smart growth approaches to land use policies and disaster resilience. The agency could implement this approach in partnership with other agencies and organizations that have expertise in smart growth approaches to development. Partners could include other state agencies, regional planning organizations, local communities, and nonprofit organizations involved in growth and development issues. These workshops could target a prioritized list of flood-prone towns and involve an evaluation of existing plans, ordinances, Flood Insurance Rate Maps, river corridor maps, drainage studies, and other relevant materials. Workshops could also evaluate each town's land use plan (if it exists) and consider if it limits public investments in flood-prone areas and encourages compact, mixed-use development in safer areas. As one example, Wisconsin hosts hazards workshops, covering topics such as applying for grants, implementing mitigation ideas, and reviewing local Hazard Mitigation Plans.[126]

- Strengthen state and local Hazard Mitigation Plans and ensure that they are coordinated with local community and land use plans. The state of California's _Community Planning and Hazard Mitigation Guidebook_ provides information about how to incorporate land use planning and climate change adaptation into local Hazard Mitigation Plans.[127,128] FEMA's _Integrating Hazard Mitigation Into Local Planning_ provides information about how to integrate hazard mitigation activities into local planning efforts.[129]

- Work with FEMA to develop improved guidance and protocols for FEMA's Public Assistance Program so that interagency interactions operate more smoothly during the next disaster. For example, one option to consider includes developing an agreed-upon protocol with FEMA to ensure that transition meetings between incoming and outgoing FEMA staff will include state and local officials, since federal staff rotations can complicate relationships with state and local agencies otherwise.

4. **Commerce, community, economic development, and housing agencies could:**

- Conduct an audit of all economic development funding decisions in the agency to determine whether they advance flood resilience goals. Examples of such funding programs include Community Development Block Grants (both pre- and post-disaster) and programs addressing community revitalization, historic preservation, tourism, business, and economic development.

- Develop a group of trained personnel who can help individuals, families, and business owners understand grant program eligibility requirements. These personnel could work in partnership with regional development corporations, small business development centers, Volunteer Organizations Active in Disasters,[130] professional associations, colleges and universities, and Community Emergency Response Team members[131] that are trained in post-disaster assistance.

For example, North Carolina's Small Business Technology Development Center provides training programs and technical assistance to help businesses prepare for and recover from disasters, including helping businesses secure federal and state disaster loans.[132]

- Conduct training programs targeting local homeowners, renters, and businesses that help to inform them about steps they can take to reduce their exposure to flood hazards and better capitalize on post-disaster recovery grant and loan programs available after disasters.

5. **Agriculture agencies could:**

- Partner with the Cooperative Extension Service[133] and Extension Disaster Education Network[134] to develop a self-assessment tool for farmers to evaluate vulnerability to floods, including steps to mitigate the impacts of flooding on individual farms and downstream neighbors, including farms, communities, and vulnerable infrastructure. The University of Florida Cooperative Extension Service provides suggestions for agricultural producers preparing for floods, including how to protect livestock during flooding events.[135] Florida's *Coastal Resilience Index* might be useful in shaping a resilience assessment tool for farmers.[136]

- Expand the role of agriculture extension agents to include hosting training programs on creating more disaster-resilient farms before the next disaster strikes. The Texas Extension Disaster Education Network provides information on disaster preparedness, mitigation, and recovery, including providing information for farmers to become better prepared for disasters.[137] Other states could provide similar information.

6. **Disaster recovery offices:**

- State governments could also consider creating and staffing a long-term flood or disaster recovery office, if one does not exist, that would be tasked with overseeing the development of a state disaster recovery plan and coordinating recovery-related policies. A disaster recovery office could lead efforts to conduct pre-disaster recovery planning, which is a core principle of the National Disaster Recovery Framework, a FEMA guide that provides a framework and flexible response structure for disaster-affected states, tribes, and local jurisdictions.[138] The Louisiana Recovery Authority, Louisiana's 33-member body tasked with identifying and obtaining funding for disaster recovery activities, is one such office.[139] The state of Iowa also created a statewide disaster recovery office, the Rebuild Iowa Office, following floods in 2008. Although the Rebuild Iowa Office closed in 2011, several of the resilience planning functions of the office continue at the state level and with the University of Iowa.[140] EPA and FEMA provided technical assistance to several communities in Iowa in partnership with the Rebuild Iowa Office and other state and local entities.[141]

5. Conclusion

While land use decisions that affect a community's flood resilience might seem to happen incrementally or opportunistically, they are often guided by plans, policies, and regulations that shape development over time. The experience from Vermont's recovery from Tropical Storm Irene suggests that coordinating local and state agency policies, plans, and actions can help promote flood recovery and encourage safer growth. This experience can serve as a model to other states and communities seeking to enhance flood resilience in the future.

The state of Vermont and communities in the Mad River Valley have already begun to implement several of the strategies outlined in this report.

State agency actions taken to date include the following:

- The Vermont Agency for Commerce and Community Development (ACCD) launched the Vermont Economic Resiliency Initiative, which will help businesses and communities continue operations and rebound quickly from future disasters. ACCD and its partners will map areas where river flooding and other hazards overlap centers of economic activity, roads, and other public investments. They will develop plans in five communities to help them better manage their risks and build back stronger and safer after disasters. These plans will serve as models for towns across the state.
- ACCD is also considering floodways when it updates certain state designation programs, including Neighborhood Development Areas and Growth Center designations.
- The Vermont Agency of Natural Resources (ANR) has hired two additional River Engineers who will assist with regulating stream alteration projects during emergencies and will deliver cross-agency training to ensure emergency actions do not exacerbate future risks. ANR is also working to improve river corridor maps and to make them more accessible to communities and organizations for assessing risks, evaluating development proposals, and identifying projects that will improve resilience to flooding.
- The Vermont Agency of Transportation (VTrans) is surveying all 60,000 culverts on state roads to map their condition and prioritize those in need of upgrades. They are working on this effort with ANR to take fluvial erosion hazards into account and to design culvert upgrades that allow for fish passage. In addition, VTrans is updating its process of prioritizing projects to include flood risk.
- The Vermont Division of Emergency Management and Homeland Security hired additional staff to work on its FEMA-funded Public Assistance Program, and it is positioned to provide support to towns in developing improved Hazard Mitigation Plans. They have also conducted a series of workshops with federal, state, and local partners to better define and improve interagency coordination.

With assistance from the Friends of the Mad River (FMR) and the Mad River Valley Planning District (MRVPD), local actions in the Mad River Valley to date include the following:

River Corridors
- After more than 2 years of planning and development, the Town of Warren passed Fluvial Erosion Hazard Zoning bylaws on November 12, 2013. FMR supported and assisted the Warren Planning Commission in community outreach and advocacy related to the bylaws. Warren joins Waitsfield, which adopted a Fluvial Erosion Hazard Overlay Zone in spring of 2011. There is more work to be done in the Towns of Fayston, Moretown, and Duxbury.

Vulnerable Settlements

- The Mad River Stream Bank Stabilization Project was completed in October 2013, which stabilizes 425 linear feet of eroding bank upstream of Waitsfield's Covered Bridge by reinforcing and riprapping the eroding bank and installing a riparian buffer. This stormwater management/flood control mitigation project at the Bridge Street Marketplace was made possible through funding from the Vermont Department of Environmental Conservation's Ecosystem Restoration Program and FEMA's Hazard Mitigation Grant Program.
- A historic building destroyed by Tropical Storm Irene located adjacent to Waitsfield's Covered Bridge was purchased by the Town of Waitsfield in January 2013 in order to restrict future development in this flood-prone location. The site is poised to become a small park.
- The Towns of Waitsfield and Moretown are in the process of moving their town offices out of vulnerable locations, both of which experienced substantial flooding from Tropical Storm Irene. These projects are made possible with support from HUD Community Development Block Grant-Disaster Recovery funding.

Safer Areas

- The Town of Waitsfield is implementing a Decentralized Wastewater Loan Program, whose pilot project is the development of wastewater capacity in a location safe from flooding.

The Whole Watershed

- To better understand the current state of stormwater management in the Mad River Valley, the FMR hired a consultant to complete a brief study entitled: *Stormwater Management Regulation in the Mad River Valley: Review and Recommendations*.[142] The goal of the study was to characterize the problems associated with stormwater in the Mad River Valley; complete a summary review of town plans and zoning regulations with respect to stormwater; and make some basic recommendations about how to improve stormwater regulations.
- To follow up on the recommendations of the stormwater study, FMR and the MRVPD met with representatives from the Planning Commissions in Warren, Waitsfield, and Fayston. FMR and MRVPD plan to continue to work with Planning Commission members to implement improved zoning regulations.
- To address existing stormwater problems, the following actions have been taken:
  - In fall 2013, the University of Vermont partnered with FMR to design and install a model bioretention facility (a type of green infrastructure) in the Village Square shopping center in Waitsfield, a priority area as identified in a recent stormwater assessment.
  - FMR is also leading a project at Mad River Glen ski area to redesign its parking lot to address stormwater issues. This project provides an excellent outreach opportunity.
  - FMR is working with Mad River Valley road crews to address areas vulnerable to erosion, including providing technical assistance and support with project development. FMR completed a Road Erosion Inventory in 2011 and is working with the towns to address priority areas.[143]

Communities across the state and nation can learn from the approaches that the state of Vermont and communities in the Mad River Valley have already implemented. The Flood Resilience Checklist in Appendix C of this report can also serve as a tool for communities to identify gaps in policies and regulations that could help improve their flood resilience. The policies, regulations, strategies, and other resources in this report (many of which are listed in Appendix D of this report) can then help communities fill those gaps and enhance their flood resilience over time.

Endnotes

[1] National Oceanic and Atmospheric Administration, National Weather Service. "Hydrologic Information Center – Flood Loss Data." http://www.nws.noaa.gov/hic. Accessed Aug. 8, 2013.

[2] Georgakakos, Aris, and Paul Fleming. *National Climate Assessment, Chapter Three: Water Resources*. 2013 (draft v. 11). http://ncadac.globalchange.gov/download/NCAJan11-2013-publicreviewdraft-chap3-water.pdf.

[3] Horton, Radley, and Gary Yohe. *National Climate Assessment, Chapter 16: Northeast*. 2013 (draft v. 11) http://ncadac.globalchange.gov/download/NCAJan11-2013-publicreviewdraft-chap16-northeast.pdf.

[4] National Research Council. *Adapting to the Impacts of Climate Change*. Washington, DC: The National Academies Press, 2010. http://www.nap.edu/openbook.php?record_id=12783.

[5] Horton, Radley, and Gary Yohe. *National Climate Assessment, Chapter 16: Northeast*. 2013 (draft v. 11) http://ncadac.globalchange.gov/download/NCAJan11-2013-publicreviewdraft-chap16-northeast.pdf.

[6] Vermont Agency of Natural Resources. "Tropical Storm Irene By the Numbers." http://www.anr.state.vt.us/anr/climatechange/irenebythenumbers.html. Accessed Feb. 6, 2014.

[7] Smart Growth Online. "Why Smart Growth?" http://www.smartgrowth.org/why.php. Accessed Jan. 23, 2014.

[8] Vermont Natural Resources Council. "Land Use Planning/Smart Growth." http://vnrc.org/programs/sustainable-communities/land-use-planningsmart-growth. Accessed Jan. 23, 2014.

[9] *Fluvial Erosion Hazard Areas: Frequently Asked Questions and Answers*. Vermont Agency of Natural Resources. 2010. http://vtwaterquality.org/rivers/docs/rv_vtfehqa.pdf.

[10] FEMA. "Multi-Hazard Mitigation Planning." http://www.fema.gov/multi-hazard-mitigation-planning. Accessed Jan. 23, 2014.

[11] FEMA. "Integrating Hazard Mitigation Into Local Planning: Case Studies and Tools for Community Officials." http://www.fema.gov/media-library/assets/documents/31372?id=7130. Accessed Apr. 8, 2014.

[12] *Integrating the Local Natural Hazard Mitigation Plan into a Community's Comprehensive Plan: A Guidebook for Local Governments*. FEMA. http://www.fema.gov/media-library-data/1388432170894-6f744a8afa8929171dc62d96da067b9a/FEMA-X-IntegratingLocalMitigation.pdf.

[13] Schwab, James. *Hazard Mitigation Planning: Integrating Best Practices into Planning*. American Planning Association, Planning Advisory Service. 2010. http://www.fema.gov/media-library-data/20130726-1739-25045-4373/pas_560_final.pdf.

[14] State of Vermont. *Vermont Statutes*. Title 24: Municipal and County Government, Chapter 117: Municipal and Regional Planning and Development, Sub-Chapter 5: Municipal Development Plan, Section 4382: The Plan for a Municipality. Effective Mar. 23, 1968. http://www.leg.state.vt.us/statutes/fullchapter.cfm?Title=24&Chapter=117.

[15] State of Rhode Island. *General Laws*. Title 45: Towns and Cities, Chapter 45-22.2: Rhode Island Comprehensive Planning and Land Use Act, Section 45-22.2-6: Required Content of a Comprehensive Plan. Revised 2011. http://webserver.rilin.state.ri.us/Statutes/TITLE45/45-22.2/45-22.2-6.HTM.

[16] *Resilient Communities Scorecard: A Tool for Assessing Your Community*. Vermont Natural Resources Council. 2013. http://vnrc.org/resources/community-planning-toolbox/tools/vermont-smart-growth-score-card.

[17] Smart Growth America. "Smart Growth Implementation Toolkit." http://www.smartgrowthamerica.org/leadership-institute/implementation-tools. Accessed Jan. 23, 2014.

[18] *Changes to the Community Rating System to Improve Disaster Resiliency and Community Sustainability*. FEMA. 2013. http://www.fema.gov/media-library-data/20130726-1907-25045-6528/changes_to_crs_system_2013.pdf.

[19] *Community Rating System*. FEMA. 2012. http://www.fema.gov/media-library-data/20130726-1605-20490-0645/communityratingsystem_2012.pdf.

[20] *Community Rating System Communities by State*. FEMA. 2012. http://www.fema.gov/media-library-data/20130726-1830-25045-0453/crosstab_bystate_4may_2012.pdf.

[21] Nelson, Kevin. *Essential Smart Growth Fixes for Rural Planning, Zoning, and Development Codes*. U.S. EPA. 2012. http://www.epa.gov/smartgrowth/essential_fixes.htm#part2.

[22] Clark, Arielle R. Farina. *Sales Tax: Earmarked for Open Space*. University of Washington. 2005. http://depts.washington.edu/open2100/pdf/3_OpenSpaceImplement/Implementation_Mechanisms/sales_tax.pdf

[23] Vermont Housing and Conservation Board. "VHCB Conservation Programs." http://www.vhcb.org/conservation.html. Accessed Mar. 6, 2014.

[24] *Implementation Manual*. Vermont Land Use Education and Training Collaborative. 2007. http://www.vpic.info/Publications/Reports/Implementation/OpenSpacePrograms.pdf.

[25] Personal communication with Faith Ingulsrud, Planning Coordinator, Vermont Department of Housing and Community Development, on Feb. 27, 2014.

[26] State of Vermont. *Vermont Statutes*. Title 24: Municipal and County Government, Chapter 77: Construction, Condemnation, Section 2804: Reserve Funds. Effective Mar. 11, 1998. http://www.leg.state.vt.us/statutes/fullsection.cfm?Title=24&Chapter=077&Section=02804.

[27] State of Vermont. *Vermont Statutes*. Title 10: Conservation and Development, Chapter 155: Acquisition of Interests in Land by Public Agencies. http://www.leg.state.vt.us/statutes/sections.cfm?Title=10&Chapter=155.

[28] Charlotte Land Trust. "Process of Conservation." http://www.charlottelandtrust.org/conservation. Accessed Apr. 3, 2014.

[29] Town of Charlotte, VT. *Select Board Meeting Notes*. August 9, 2010. http://www.charlottevt.org/vertical/sites/%7B5618C1B5-BAB5-4588-B4CF-330F32AA3E59%7D/uploads/%7B56035679-59BD-419A-8D4E-7D23A177FD50%7D.PDF.

[30] State of Vermont. *Vermont Statutes*. Title 24: Municipal and County Government, Chapter 131: Impact Fees, Section 5200: Purpose. Effective Jul. 1, 1989. http://www.leg.state.vt.us/statutes/fullsection.cfm?Title=24&Chapter=131&Section=05200.

[31] The Vermont River Conservancy. "Completed Project List." http://www.vermontriverconservancy.org/completed-projects/list. Accessed Mar. 6, 2014.

[32] FEMA. "Hazard Mitigation Assistance – Property Acquisition (Buyouts)." http://www.fema.gov/application-development-process/hazard-mitigation-assistance-property-acquisition-buyouts. Accessed Jan. 23, 2014.

[33] *Property Acquisition Handbook for Local Communities: A Summary for States*. FEMA. 1998. http://www.fema.gov/media-library/assets/documents/3117.

[34] *2013 National Award for Smart Growth Achievement*. U.S. EPA. 2013. http://www.epa.gov/smartgrowth/awards/sg_awards_publication_2013.htm#plazas.

[35] State of Vermont. *Vermont Statutes*. Title 24: Municipal and County Government, Chapter 117: Municipal and Regional Planning and Development, Section 4423: Transfer of Development Rights. Amended 2003. http://www.leg.state.vt.us/statutes/fullsection.cfm?Title=24&Chapter=117&Section=04423.

[36] *TDR Program Overview*. Department of Economic Development, Agricultural Services Division. 2006. http://www6.montgomerycountymd.gov/content/ded/agservices/pdffiles/tdr_info.pdf.

[37] Pinho, Rute. *Maryland's Transfer of Development Rights Programs*. 2010. http://www.cga.ct.gov/2010/rpt/2010-R-0464.htm.

[38] *The New Jersey Pinelands Development Credit (PDC) Program*. New Jersey Pinelands Commission. 2012. http://www.nj.gov/pinelands/infor/fact/PDCfacts.pdf.

[39] Commonwealth of Virginia. *Riparian forest buffer protection for waterways tax credit*. Section 58.1-339.10. http://leg1.state.va.us/cgi-bin/legp504.exe?000+cod+58.1-339.10.

[40] Vermont Department of Taxes. "Current Use." http://www.state.vt.us/tax/pvrcurrentuse.shtml. Accessed Apr. 3, 2014.

[41] Duerksen, Chris, and Cara Snyder. *Nature Friendly Communities: Habitat Protection and Land Use Planning*. Washington, D.C.: Island Press, 2005.

[42] U.S. Department of Agriculture. "Conservation Reserve Enhancement Program – Vermont." http://www.fsa.usda.gov/FSA/newsReleases?area=newsroom&subject=landing&topic=pfs&newstype=prfactsheet&type=detail&item=pf_20110214_consv_en_crepvt01.html. Accessed Apr. 3, 2014.

[43] U.S. Department of Agriculture. "Conservation." http://www.usda.gov/wps/portal/usda/usdahome?navid=CONSERVATION. Accessed Apr. 3, 2014.

[44] *Ready Ag: Disaster and Defense Preparedness for Production Agriculture.* Penn State Cooperative Extension. 2010. http://readyag.psu.edu/pdfs/ReadyAG_DAIRYandGENERALWorkbook.pdf.

[45] Extension Disaster Education Network. "Reducing the Impact of Disasters Through Education." http://eden.lsu.edu/Pages/default.aspx. Accessed Apr. 3, 2014.

[46] U.S. Government. *Code of Federal Regulations.* Title 44: Emergency Management and Assistance, Chapter 1: Federal Emergency Management Agency, Department of Homeland Security, Subchapter B: Insurance and Hazard Mitigation, Part 59: General Provisions. http://www.ecfr.gov/cgi-bin/text-idx?tpl=/ecfrbrowse/Title44/44cfr59_main_02.tpl.

[47] U.S. Government. *Code of Federal Regulations.* Title 44: Emergency Management and Assistance, Chapter 1: Federal Emergency Management Agency, Department of Homeland Security, Subchapter B: Insurance and Hazard Mitigation, Part 60: Criteria for Land Management and Use. http://www.ecfr.gov/cgi-bin/text-idx?tpl=/ecfrbrowse/Title44/44cfr60_main_02.tpl.

[48] Association of State Flood Plain Managers. "No Adverse Impact Floodplain Management." http://www.floods.org/index.asp?menuID=349&fir. Accessed Apr. 3, 2014.

[49] FEMA. "Definitions." http://www.fema.gov/national-flood-insurance-program/definitions#F. Accessed Apr. 8, 2014.

[50] FEMA. "Floodway." http://www.fema.gov/floodplain-management/floodway. Accessed Apr. 8, 2014.

[51] Dolan, Kari, and Mike Kline. *Municipal Guide to Fluvial Erosion Hazard Mitigation.* Vermont Agency of Natural Resources. 2011. http://www.vtwaterquality.org/rivers/docs/rv_municipalguide.pdf.

[52] Stearns County, MN. "Zoning Districts." http://www.co.stearns.mn.us/Environment/LandUseandSubdivision/Zoning/ZoningDistricts. Accessed Apr. 3, 2014.

[53] Minnesota Department of Agriculture. "Minnesota's Agricultural Land Preservation Statutes." http://www.mda.state.mn.us/protecting/conservation/preservation/statutes.aspx. Accessed Apr. 3, 2014.

[54] Blaine County, ID. *County Code, Title 9, Chapter 6B: Resource Conservation District (RC-160).* Passed Mar. 19, 2013. http://www.sterlingcodifiers.com/codebook/index.php?book_id=450.

[55] Town of Colchester, VT. *Zoning Regulations, Table A-2 Dimensional Standards.* Amended Dec. 10, 2013. http://colchestervt.gov/PlanningZ/regs/Zoning/A-2-DimensionalReq.pdf.

[56] Town of Windsor, VT. *Zoning Regulations.* Amended Sept. 25, 2007. http://swcrpc.org/wp-content/uploads/2013/11/Windsor-Zoning-Regulations-2007.pdf.

[57] Town of Windsor, VT. *Subdivision Regulations.* Amended Sept. 26, 2006. http://swcrpc.org/wp-content/uploads/2013/11/Windsor-Subdivision-Regulations-2006.pdf.

[58] Town of Hartford, VT. *Ordinances.* Jul. 25, 2013. https://law.resource.org/pub/us/code/city/vt/Hartford.html#13455808.

[59] Town of St. George, VT. *Land Use Regulations.* Adopted Jul. 22, 2010. http://www.stgeorgevt.com/pdfs/Regulations%20and%20Bylaws/SGLUR(22Jul2010)lr.pdf.

[60] Nelson, Kevin. *Essential Smart Growth Fixes for Rural Planning, Zoning, and Development Codes.* U.S. EPA. 2012. http://www.epa.gov/smartgrowth/essential_fixes.htm#part2.

[61] FEMA. "Flood Insurance Reform." http://www.fema.gov/flood-insurance-reform. Accessed Apr. 3, 2014.

[62] *Engineering with Nature: Alternative Techniques to Riprap Bank Stabilization.* FEMA. http://www.fema.gov/pdf/about/regions/regionx/Engineering_With_Nature_Web.pdf.

[63] U.S. Army Corps of Engineers. "Mission Overview." http://www.usace.army.mil/Missions.aspx. Accessed Apr. 8, 2014.

[64] FEMA. "Public Assistance: Local, State, Tribal and Non-Profit." http://www.fema.gov/public-assistance-local-state-tribal-and-non-profit. Accessed Apr. 8, 2014.

[65] FEMA. "Pre-Disaster Mitigation Grant Program." http://www.fema.gov/pre-disaster-mitigation-grant-program. Accessed Apr. 8, 2014.

[66] FEMA. "Hazard Mitigation Grant Program." http://www.fema.gov/hazard-mitigation-grant-program. Accessed Apr. 8, 2014.

[67] FEMA. "Response and Recovery." http://www.fema.gov/response-recovery. Accessed Apr. 8, 2014.

[68] U.S. Department of Housing and Urban Development. "State Administered Community Development Block Grant." http://portal.hud.gov/hudportal/HUD?src=/program_offices/comm_planning/communitydevelopment/programs/stateadmin. Accessed Apr. 8, 2014.

[69] U.S. Department of Housing and Urban Development. "Community Development Block Grant Program – CDBG." http://portal.hud.gov/hudportal/HUD?src=/program_offices/comm_planning/communitydevelopment/programs. Accessed Apr. 8, 2014.

[70] Vermont Agency of Commerce and Community Development. "Community Development Block Grant Disaster Recovery Funds." http://accd.vermont.gov/strong_communities/opportunities/funding/cdbgdr. Accessed Apr. 8, 2014.

[71] U.S. Department of Housing and Urban Development. "Community Development Block Grant Entitlement Communities Grants." http://portal.hud.gov/hudportal/HUD?src=/program_offices/comm_planning/communitydevelopment/programs/entitlement#eligibleactivities. Accessed Apr. 8, 2014.

[72] *Emergency Relief Manual.* Federal Highway Administration. 2013. http://www.fhwa.dot.gov/reports/erm/er.pdf.

[73] FEMA. "Base Flood Elevation." http://www.fema.gov/national-flood-insurance-program/base-flood-elevation. Accessed Feb. 27, 2014.

[74] *No Adverse Impact Floodplain Management Community Case Studies.* Association of State Flood Plain Managers. 2004. http://www.floods.org/PDF/NAI_Case_Studies.pdf.

[75] *No Adverse Impact Status Report: Helping Communities Implement NAI.* Association of State Flood Plain Managers. 2002. http://www.floods.org/NoAdverseImpact/NAI_Status_Report.pdf.

[76] Williams, Eric. *Innovative Land Use Planning Techniques: A Handbook for Sustainable Development.* Chapter 2.7 Flood Hazard Area Zoning. New Hampshire Department of Environmental Services. 2008. http://des.nh.gov/organization/divisions/water/wmb/repp/documents/ilupt_complete_handbook.pdf.

[77] Georgakakos, Aris, and Paul Fleming. *National Climate Assessment, Chapter Three: Water Resources.* 2013 (draft v. 11). http://ncadac.globalchange.gov/download/NCAJan11-2013-publicreviewdraft-chap3-water.pdf.

[78] Medlock, Sam Riley. *Preparing for the Next Flood: Vermont Floodplain Management.* Land Use Institute, Vermont Law School. 2009. http://www.vermontlaw.edu/Documents/VLS.065.09%20LAND%20USE%20PAPER_PFF.pdf.

[79] Personal communication with John E. Adams, Planning Coordinator, Vermont Department of Housing and Community Development, on Mar. 11, 2014.

[80] Vermont Department of Public Safety. "Code Information Sheets." http://firesafety.vermont.gov/resources/code_sheets. Accessed Mar. 13, 2014.

[81] *Code Information Sheet: Permit and Licensing Requirements.* Vermont Department of Public Safety. http://firesafety.vermont.gov/sites/firesafety/files/pdf/Code%20Info%20Sheets/2012%20permit%20requirements.pdf. Accessed Mar. 13, 2014.

[82] International Code Council. "International Code Council." http://www.iccsafe.org. Accessed Apr. 8, 2014.

[83] FEMA. "Highlights of ASCE 24-05, Flood Resistant Design and Construction (2010)." http://www.fema.gov/library/viewRecord.do?id=3515. Accessed Apr. 8, 2014.

[84] International Code Council. "International Green Construction Code." http://www.iccsafe.org/cs/IGCC. Accessed Apr. 8, 2014.

[85] *Smart Growth for Coastal and Waterfront Communities*. National Oceanic and Atmospheric Administration, U.S. EPA, ICMA, and Sea Grant Rhode Island. 2009. http://coastalsmartgrowth.noaa.gov/smartgrowth_fullreport.pdf.

[86] Napa County, CA. "Flood Control and Water Conservation District." http://www.countyofnapa.org/Pages/DepartmentContent.aspx?id=4294971816. Accessed Apr. 8, 2014.

[87] Environmental Finance Center. "Environmental Finance Center." http://www.efc.umd.edu. Accessed Apr. 8, 2014.

[88] Personal communication with Faith Ingulsrud, Planning Coordinator, Vermont Department of Housing and Community Development, on Mar. 7, 2014.

[89] Vermont Agency of Natural Resources. "Flood Resilience." https://outside.vermont.gov/agency/ANR/FloodResilience/Pages/default.aspx. Accessed Mar. 13, 2014.

[90] Chittenden County, VT. "Smarter WaterWays." http://www.smartwaterways.org. Accessed Mar. 6, 2014.

[91] Metropolitan North Georgia Water Planning District. "Watershed Management Plan." http://www.northgeorgiawater.com/plans/watershed-management-plan. Accessed Apr. 8, 2014.

[92] *The Citizen's Guide to Stormwater Pollution Prevention*. City of Arlington, Texas. 2010. http://www.arlingtontx.gov/environmentalservices/pdf/CitizensStormwaterGuide.pdf.

[93] *Funding Stormwater Programs*. U.S. EPA. 2009. http://www.epa.gov/region1/npdes/stormwater/assets/pdfs/FundingStormwater.pdf.

[94] Vermont League of Cities and Towns. *Vermont Town Road and Bridge Standards*, Culverts and Bridges. 2013. http://www.vlct.org/assets/News/Current/Town_Road_Bridge_Standards.pdf.

[95] State of Vermont. *Vermont Statutes*. Title 10: Conservation and Development, Chapter 111: Fish, Section 4607: Obstructing Streams. Effective May 9, 1961. http://www.leg.state.vt.us/statutes/fullsection.cfm?Title=10&Chapter=111&Section=04607.

[96] Bates, Kozmo Ken, and Rich Kirn. *Guidelines for the Design of Stream/Road Crossings for Passage of Aquatic Organisms in Vermont*. Vermont Department of Fish and Wildlife. 2009. http://www.vtfishandwildlife.com/library/Reports_and_Documents/Aquatic%20Organism%20Passage%20at%20Stream%20Crossings/_Guidelines%20for%20the%20Design%20of%20Stream_Road%20Crossings%20for%20Passage%20of%20Aquatic%20Organisms%20in%20Vermont.pdf.

[97] Town of Williston, VT. *Unified Development Bylaw*. Chapter 29, Watershed Health. Amended Jul. 19, 2010. http://town.williston.vt.us/vertical/Sites/%7BF506B13C-605B-4878-8062-87E5927E49F0%7D/uploads/%7B2920AC61-60E4-483B-8A02-015028396045%7D.PDF.

[98] Williston, VT. "Stormwater." http://www.town.williston.vt.us/index.asp?Type=B_BASIC&SEC=%7BACC6B21E-0FDB-497F-8A5A-62CDFF871272%7D. Accessed April 3, 2014.

[99] *Stormwater Management Plan*. Town of Rutland, VT. 2013. http://www.vtwaterquality.org/stormwater/docs/ms4/sw_TownofRutland_MS4_SWMP.pdf.

[100] *Stormwater Management Plan*. City of South Burlington, VT. 2013. http://www.watershedmanagement.vt.gov/stormwater/docs/ms4/sw_SBurlington_SWMP.pdf.

[101] U.S. EPA "Why Green Infrastructure?" http://water.epa.gov/infrastructure/greeninfrastructure/gi_why.cfm. Accessed Apr. 8, 2014.

[102] U.S. EPA. "Stormwater Management Best Practices." http://www.epa.gov/oaintrnt/stormwater/best_practices.htm. Accessed Apr. 8, 2014.

[103] New York City Department of Environmental Protection. "Blue Roof and Green Roof." http://www.nyc.gov/html/dep/html/stormwater/green_pilot_project_ps118.shtml. Accessed Apr. 9, 2014.

[104] Town of Williston, VT. *Unified Development Bylaw*. Chapter 29, Watershed Health. Amended Jul. 19, 2010. http://town.williston.vt.us/vertical/Sites/%7BF506B13C-605B-4878-8062-87E5927E49F0%7D/uploads/%7B2920AC61-60E4-483B-8A02-015028396045%7D.PDF.

[105] Duerksen, Chris. *Tree Conservation Ordinances: Land-Use Regulations Go Green*. American Planning Association. 1993.

[106] Currituck County, NC. *Unified Development Ordinance*. Section 7.2, Tree Protection. Amended Nov. 18, 2013. http://co.currituck.nc.us/pdf/unified-development-ordinance-new/Currituck%20UDO%20Final%20-%2011-19-2013red.pdf.

[107] Folly Beach, SC. *Code of Ordinances*. Title XV, Section 166.01, Tree Protection. Passed Jul. 23, 2013. http://www.amlegal.com/nxt/gateway.dll/South%20Carolina/follybeach/follybeachsouthcarolinacodeofordinances?f=templates$fn=default.htm$3.0$vid=amlegal:follybeach_sc.

[108] Town of Wellesley MA. *Rules and Regulations Relative to the Administration of Section XVIE: Tree Preservation and Protection*. http://www.wellesleyma.gov/Pages/WellesleyMA_Planning/TreeBylawRulesRegs6.27.11.pdf.

[109] Town of Williston, VT. *Unified Development Bylaw*. Chapter 29, Watershed Health. Amended Jul. 19, 2010. http://town.williston.vt.us/vertical/Sites/%7BF506B13C-605B-4878-8062-87E5927E49F0%7D/uploads/%7B2920AC61-60E4-483B-8A02-015028396045%7D.PDF.

[110] Salt Lake County, UT. *Code of Ordinances*. Title 19: Zoning, Chapter 19.72: Foothills and Canyons Overlay Zone. http://library.municode.com/HTML/16602/level2/TIT19ZO_CH19.72FOCAOVZO.html.

[111] Mitchell, Paul. *The Scientific Justification for Stream Buffers*. University of Georgia Land Use Clinic. 2006. http://www.rivercenter.uga.edu/publications/pdf/luc_buffer_fact_sheet.pdf. Accessed Apr. 9, 2014.

[112] Smith, Gavin, Dylan Sandler, and Mikey Goralnik. *Vermont State Agency Policy Options: Smart Growth Implementation Assistance Program, Disaster Recovery and Long-Term Resilience Planning in Vermont*. U.S. Department of Homeland Security Coastal Hazards Center of Excellence, University of North Carolina at Chapel Hill. http://accd.vermont.gov/sites/accd/files/Documents/strongcommunities/cpr/VT-StateAgencyPolicyOptionsFINAL_web.pdf.

[113] Smith, Gavin, Dylan Sandler, and Mikey Goralnik. *Vermont State Agency Policy Options: Smart Growth Implementation Assistance Program, Disaster Recovery and Long-Term Resilience Planning in Vermont*. U.S. Department of Homeland Security Coastal Hazards Center of Excellence, University of North Carolina at Chapel Hill. http://accd.vermont.gov/sites/accd/files/Documents/strongcommunities/cpr/VT-StateAgencyPolicyOptionsFINAL_web.pdf.

[114] Smith, Gavin, Dylan Sandler, and Mikey Goralnik. "Assessing State Policy Linking Disaster Recovery, Smart Growth, and Resilience in Vermont Following Tropical Storm Irene." *Vermont Journal of Environmental Law*. Vol. 15 (2013). 66-102. http://vjel.vermontlaw.edu/files/2013/11/Smith.pdf.

[115] Smith, Gavin, and Dylan Sandler. *State Disaster Recovery Planning Guide*. U.S. Department of Homeland Security Coastal Hazards Center of Excellence, University of North Carolina at Chapel Hill. 2012. http://coastalhazardscenter.org/dev/wp-content/uploads/2012/05/State-Disaster-Recovery-Planning-Guide_2012.pdf.

[116] FEMA. "Community Planning and Capacity Building." http://www.fema.gov/community-planning-and-capacity-building. Accessed Apr. 8, 2014.

[117] Hillsborough County, FL. "Post-Disaster Redevelopment Plan Documents." http://www.hillsboroughcounty.org/index.aspx?nid=1795. Accessed Apr. 8, 2014.

[118] Smith, Gavin. *Planning for Post-Disaster Recovery: A Review of the United States Disaster Assistance Framework*. Gavin Smith. Island Press, 2012.

[119] *No Adverse Impact Status Report: Helping Communities Implement NAI*. Association of State Flood Plain Managers. 2002. http://www.floods.org/NoAdverseImpact/NAI_Status_Report.pdf.

[120] Schwab, James C. *Hazard Mitigation: Integrating Best Practices into Planning*. American Planning Association. 2010. Pages 74-86. http://www.fema.gov/media-library-data/20130726-1739-25045-4373/pas_560_final.pdf.

[121] Town of Bennington, VT. *Fluvial Erosion Hazard Overlay District*. Adopted Apr., 27, 2009. http://www.benningtonplanningandpermits.com/BPC/wp-content/uploads/2011/02/fehr.pdf.

[122] Association of State Flood Plain Managers. "No Adverse Impact." http://www.floods.org/index.asp?menuID=460. Accessed Apr. 8, 2014.

[123] Edwards, Frances L. and Daniel C. Goodrich. *Handbook of Emergency Management for State-Level Transportation Agencies*. San Jose State University. 2010. http://transweb.sjsu.edu/MTIportal/research/publications/documents/COOP%20COG%20I_Vince_022410.pdf.

[124] *Culvert Inventory and Inspection Manual*. New York State Department of Transportation. 2006. https://www.dot.ny.gov/divisions/operating/oom/transportation-maintenance/repository/CulvertInventoryInspectionManual.pdf.

[125] *Culvert Management Manual.* Ohio Department of Transportation. 2014. http://www.dot.state.oh.us/Divisions/Engineering/Hydraulics/Culvert%20Management/Culvert%20Management%20Manual/CMM%20-%20January2014.pdf.

[126] Wisconsin Department of Military Affairs, Division of Emergency Management. "2012 All-Hazards Mitigation Planning Workshop Presentations and Handouts." http://emergencymanagement.wi.gov/mitigation/Mitigation_Workshop/toc.asp. Accessed Apr. 8, 2014.

[127] California Governor's Office of Emergency Services. "Hazard Mitigation." http://www.calema.ca.gov/hazardmitigation. Accessed Apr. 8, 2014.

[128] California Emergency Management Agency. "Local Hazard Mitigation Planning Program (LHMP)." http://hazardmitigation.calema.ca.gov/plan/local_hazard_mitigation_plan_lhmp. Accessed Apr. 8, 2014.

[129] FEMA. "Integrating Hazard Mitigation Into Local Planning: Case Studies and Tools for Community Officials." http://www.fema.gov/media-library/assets/documents/31372?id=7130. Accessed Apr. 8, 2014.

[130] National Voluntary Organizations Active in Disaster. http://www.nvoad.org. Accessed Apr. 8, 2014.

[131] FEMA. "Community Emergency Response Teams." https://www.fema.gov/community-emergency-response-teams. Accessed May 12, 2014.

[132] Small Business Technology Development Center. http://www.sbtdc.org. Accessed Apr. 8, 2014.

[133] U.S. Department of Agriculture National Institute of Food and Agriculture. "Cooperative Extension Offices." http://www.csrees.usda.gov/Extension. Accessed Apr. 8, 2014.

[134] Extension Disaster Education Network. http://eden.lsu.edu/Pages/default.aspx. Accessed Apr. 8, 2014.

[135] *Special Considerations for Agricultural Producers-Preparing for a Flood or a Flash Flood*. University of Florida Cooperative Extension Service. 1998. http://disaster.ifas.ufl.edu/PDFS/CHAP09/D09-07.pdf.

[136] Sempier, T.T., et al. *Coastal Resilience Index: A Community Self-Assessment*. Mississippi-Alabama Sea Grant Consortium and National Oceanic and Atmospheric Administration. 2010. http://www.southernclimate.org/documents/resources/Coastal_Resilience_Index_Sea_Grant.pdf.

[137] Texas A&M AgriLife Extension. "Texas Extension Disaster Education Network." http://texashelp.tamu.edu. Accessed Apr. 8, 2014.

[138] FEMA. "National Disaster Recovery Framework." http://www.fema.gov/national-disaster-recovery-framework. Accessed Apr. 8, 2014.

[139] *Louisiana Recovery Authority Strategic Plan: FY 2008/2009*. Louisiana Recovery Authority. http://lra.louisiana.gov/assets/docs/searchable/StrategicPlan0809.pdf.

[140] University of Iowa School of Urban and Regional Planning. "RIO Iowa Project." http://rio.urban.uiowa.edu. Accessed Apr. 9, 2014.

[141] U.S. EPA. "Smart Growth Technical Assistance in Iowa." http://www.epa.gov/smartgrowth/iowa_techasst.htm. Accessed Apr. 8, 2014.

[142] *Stormwater Management Regulation in the Mad River Valley: Review and Recommendations*. Watershed Consulting Associates, LLC. 2013. http://www.friendsofthemadriver.org/documents/MRVStormwater_Scoping_Study_Spring_2013_.pdf.

[143] Mad River Valley Erosion Study Final Report. Watershed Consulting Associates, LLC. 2012. http://friendsofthemadriver.org/documents/MRV_Road_Erosion_Study_Report.pdf.

Appendix A: About the Environmental Protection Agency's Smart Growth Implementation Assistance Program

Communities around the country are looking to get the most from new development and to maximize their investments. Frustrated by development that gives residents no choice but to drive long distances between jobs and housing, many communities are bringing workplaces, homes, and services closer together. Communities are examining and changing zoning codes that make it impossible to build neighborhoods with a variety of housing types. They are questioning the fiscal wisdom of neglecting existing infrastructure while expanding new sewers, roads, and services into the fringe. Many places that have been successful in ensuring that development improves their community, economy, and environment have used smart growth principles to do so (see box). Smart growth describes development patterns that create attractive, distinctive, and walkable communities that give people of varying age, wealth, and physical ability a range of safe, convenient choices in where they live and how they get around. Growing smart also means that we use our existing resources efficiently and preserve the lands, buildings, and environmental features that shape our neighborhoods, towns, and cities.

Smart Growth Principles

Based on the experience of communities around the nation, the Smart Growth Network developed a set of 10 basic principles:

- Mix land uses.
- Take advantage of compact building design.
- Create a range of housing opportunities and choices.
- Create walkable neighborhoods.
- Foster distinctive, attractive communities with a strong sense of place.
- Preserve open space, farmland, natural beauty, and critical environmental areas.
- Strengthen and direct development towards existing communities.
- Provide a variety of transportation choices.
- Make development decisions predictable, fair, and cost effective.
- Encourage community and stakeholder collaboration in development decisions.

Source: Smart Growth Network. "Why Smart Growth?" www.smartgrowth.org/why.php

However, communities often need additional tools, resources, or information to achieve these goals. In response to this need, the U.S. Environmental Protection Agency (EPA) launched the Smart Growth Implementation Assistance (SGIA) Program to provide technical assistance—through contractor services—to selected communities.

The goals of this assistance are to improve the overall climate for infill, brownfields redevelopment, and the revitalization of non-brownfield sites—as well as to promote development that meets economic, community, public health, and environmental goals. EPA and its contractors assemble teams whose members have expertise that meets community needs. While engaging community participants on their aspirations for development, the team can bring their experiences from working in other parts of the country to provide best practices for the community to consider.

For more information on the SGIA program, including reports from communities that have received assistance, see *www.epa.gov/smartgrowth/sgia.htm*.

Appendix B: About the Project

This appendix describes the process by which the state and local assessments for this Smart Growth Implementation Assistance project were completed.

A. Local Policy Assessment

The local policy assessment, funded by the U.S. Environmental Protection Agency (EPA) and completed by consultants from SRA International, Inc., Clarion Associates, and CSA Ocean Sciences, Inc., included the steps listed below. Communities seeking to improve their flood resilience may wish to consider these steps.

1. **Identify and review plans, policies, codes, and regulations that affect flood resilience.**

    A team of national experts in hazard mitigation, flood recovery, land use planning, and state policy worked with officials from the state of Vermont, regional planning organizations, and local municipalities to discuss flood history, flood damage, and development and demographic trends in the Mad River Valley and to identify key documents for the team to review for the communities of Moretown and Waitsfield. Moretown and Waitsfield were chosen because they were representative of other Vermont communities affected by Tropical Storm Irene. The team reviewed Moretown and Waitsfield's codes, their local Hazard Mitigation Plans, the regional land use plan that covered both towns,

> ### Key Planning Documents for Flood Resilience Review
>
> In most jurisdictions, the primary documents that the community would review for flood resilience include:
>
> - Local comprehensive plans.
> - Local Hazard Mitigation Plans.
> - Zoning and subdivision regulations (including flood plain development standards).
> - Building codes.
> - Stormwater management ordinances.
> - Regional plans.
>
> Smaller towns and villages may not have stand-alone building codes and instead might include building code-type regulations in local zoning or subdivision regulations. Likewise, if the town does not have a comprehensive stormwater management ordinance, some aspects of stormwater management can be addressed in zoning and subdivision regulations or by standards established by state environmental or natural resource agencies.

and other relevant policies. Because Moretown and Waitsfield did not have building codes, the team reviewed the zoning and subdivision provisions addressing building code issues.

The team then developed a framework for reviewing the documents (which eventually became the checklist in Appendix C). The initial assessment was organized into three general categories representing the range of options that communities can typically use to achieve safer growth:

- Protect undeveloped river corridors, including vulnerable areas, such as flood plains and wetlands along waterways, from incompatible development.

- Protect people, buildings, and facilities in already-developed, vulnerable areas.

- Encourage new development in safer areas.

For each category, the team identified specific policies, regulations, or non-regulatory approaches that other jurisdictions have used successfully and then determined whether Moretown or Waitsfield had used those approaches. For example, in the category of protecting undeveloped river corridors, the team assessed whether zoning regulations addressed development on steep slopes or included stream buffer standards. In the category of protecting people and buildings in already–

developed, vulnerable areas, the team assessed whether current zoning regulations would protect structures that are rebuilt.

2. Develop initial policy and regulatory options.

Based on the review of policies, the team prepared a detailed assessment that identified a range of policy options and implementation tools, both regulatory and non-regulatory, that the two Mad River Valley towns might consider to improve their flood resilience. These initial policy options were distributed to state, regional, and local officials prior to the team's site visit to the Mad River Valley.

On October 23-25, 2012, the team, including federal and state officials, visited the Mad River Valley to view the extent of flood damage and discuss the initial policy options with stakeholders. During this visit, the team met with town officials in Waitsfield and Moretown, including the zoning administrator, town manager, elected and appointed officials for each town, and representatives from regional planning and nonprofit organizations to discuss the policy options and receive feedback. The site visit also included a community meeting during which the team presented the policy options to residents, business owners, local officials, and other stakeholders from the Mad River Valley and solicited feedback on those ideas.

3. Refine the checklist and policy and regulatory options.

Based on the input gathered during the site visit, the team revised the flood resilience checklist and policy options to improve flood resilience in the Mad River Valley. The team organized these policy options into four geographically oriented approaches, adapted from the original three categories:

- **River Corridors**: Conserve land and discourage development in particularly vulnerable areas along river corridors such as flood plains and wetlands.

Figure B-1. In October 2012, EPA, FEMA, and Vermont state agency staff toured flood-damaged sites in the Mad River Valley. Credit: EPA.

- **Vulnerable Settlements**: Where development already exists in vulnerable areas, protect people, buildings, and facilities to reduce future flooding risk.

- **Safer Areas**: Plan for and encourage new development in areas that are less vulnerable to future flooding events.

- **The Whole Watershed:** Implement enhanced stormwater management techniques to slow, spread, and infiltrate floodwater.

These policy options, summarized in this report, are described in more detail in a policy memo for Moretown and Waitsfield and a guidance document for the state of Vermont, available at: *http://accd.vermont.gov/strong_communities/opportunities/planning/resiliency/sgia*.

B. State Policy Assessment

The state policy assessment, led by faculty and staff from the University of North Carolina at Chapel Hill's Department of Homeland Security Coastal Hazards Center of Excellence (the Coastal Hazards Center team) and funded by the Federal Emergency Management Agency (FEMA), followed a parallel process:

1. Analyze state policies from a flood resilience perspective.

The Coastal Hazards Center team analyzed relevant state policies from a variety of state-level organizations in Vermont, including the Agency of Natural Resources; Agency of Transportation; Division of Emergency Management and Homeland Security; Agency of Commerce and Community Development; Agency of Agriculture, Food, and Markets; and the Irene Recovery Office. The team assessed these agencies' policies in terms of their ability to encourage flood resilience at the local level.

2. Participate in a site visit to the Mad River Valley.

The Coastal Hazards Center team participated in the October 2012 site visit to the Mad River Valley. During the visit, the team talked with state agency officials to learn how state activities and policies might influence flood resilience at the local level, both in the Mad River Valley communities that were the focus of this project and in other communities throughout the state.

3. Draft, review, and finalize policy options for state-level organizations.

Following the site visit, the Coastal Hazards Center team drafted initial policy options and presented these policy options to state agency representatives at a follow-up meeting on July 24, 2013. After this meeting, the team refined and finalized a memo on policy options for Vermont agencies to consider and delivered it to the state agencies.

A detailed report on the state policy assessment and suggested policy options is available at *http://accd.vermont.gov/strong_communities/opportunities/planning/resiliency/sgia*. Some material from that report is included in Section 4 of this document.

Appendix C: Flood Resilience Checklist

Is your community prepared for a possible flood? Completing this flood resilience checklist can help you begin to answer that question.

What is the Flood Resilience Checklist?

This checklist includes overall strategies to improve flood resilience as well as specific strategies to conserve land and discourage development in river corridors; to protect people, businesses, and facilities in vulnerable settlements; to direct development to safer areas; and to implement and coordinate stormwater management practices throughout the whole watershed.

Who should use it?

This checklist can help communities identify opportunities to improve their resilience to future floods through policy and regulatory tools, including comprehensive plans, Hazard Mitigation Plans, local land use codes and regulations, and non-regulatory programs implemented at the local level. Local government departments such as community planning, public works, and emergency services; elected and appointed local officials; and other community organizations and nonprofits can use the checklist to assess their community's readiness to prepare for, deal with, and recover from floods.

Why is it important?

Completing this checklist is the first step is assessing how well a community is positioned to avoid and/or reduce flood damage and to recover from floods. If a community is not yet using some of the strategies listed in the checklist and would like to, the policy options and resources listed in this report can provide ideas for how to begin implementing these approaches.

FLOOD RESILIENCE CHECKLIST		
Overall Strategies to Enhance Flood Resilience (Learn more in Section 2, pp. 9-11)		
1. Does the community's comprehensive plan have a hazard element or flood planning section?	☐ Yes	☐ No
a. Does the comprehensive plan cross-reference the local Hazard Mitigation Plan and any disaster recovery plans?	☐ Yes	☐ No
b. Does the comprehensive plan identify flood- and erosion-prone areas, including river corridor and fluvial erosion hazard areas, if applicable?	☐ Yes	☐ No
c. Did the local government emergency response personnel, flood plain manager, and department of public works participate in developing/updating the comprehensive plan?	☐ Yes	☐ No
2. Does the community have a local Hazard Mitigation Plan approved by the Federal Emergency Management Agency (FEMA) and the state emergency management agency?	☐ Yes	☐ No
a. Does the Hazard Mitigation Plan cross-reference the local comprehensive plan?	☐ Yes	☐ No

FLOOD RESILIENCE CHECKLIST		
b. Was the local government planner or zoning administrator involved in developing/updating the Hazard Mitigation Plan?	☐ Yes	☐ No
c. Were groups such as local businesses, schools, hospitals/medical facilities, agricultural landowners, and others who could be affected by floods involved in the Hazard Mitigation Plan drafting process?	☐ Yes	☐ No
d. Were other local governments in the watershed involved to coordinate responses and strategies?	☐ Yes	☐ No
e. Does the Hazard Mitigation Plan emphasize non-structural pre-disaster mitigation measures such as acquiring flood-prone lands and adopting No Adverse Impact flood plain regulations?	☐ Yes	☐ No
f. Does the Hazard Mitigation Plan encourage using green infrastructure techniques to help prevent flooding?	☐ Yes	☐ No
g. Does the Hazard Mitigation Plan identify projects that could be included in pre-disaster grant applications and does it expedite the application process for post-disaster Hazard Mitigation Grant Program acquisitions?	☐ Yes	☐ No
3. Do other community plans (e.g., open space or parks plans) require or encourage green infrastructure techniques?	☐ Yes	☐ No
4. Do all community plans consider possible impacts of climate change on areas that are likely to be flooded?	☐ Yes	☐ No
5. Are structural flood mitigation approaches (such as repairing bridges, culverts, and levees) and non-structural approaches (such as green infrastructure) that require significant investment of resources coordinated with local capital improvement plans and prioritized in the budget?	☐ Yes	☐ No
6. Does the community participate in the National Flood Insurance Program Community Rating System?	☐ Yes	☐ No
Conserve Land and Discourage Development in River Corridors (Learn more in Section 3.A, pp. 14-19)		
1. Has the community implemented non-regulatory strategies to conserve land in river corridors, such as:		
a. Acquisition of land (or conservation easements on land) to allow for stormwater absorption, river channel adjustment, or other flood resilience benefits?	☐ Yes	☐ No
b. Buyouts of properties that are frequently flooded?	☐ Yes	☐ No
c. Transfer of development rights program that targets flood-prone areas as sending areas and safer areas as receiving areas?	☐ Yes	☐ No
d. Tax incentives for conserving vulnerable land?	☐ Yes	☐ No

FLOOD RESILIENCE CHECKLIST		
e. Incentives for restoring riparian and wetland vegetation in areas subject to erosion and flooding?	☐ Yes	☐ No
2. Has the community encouraged agricultural and other landowners to implement pre-disaster mitigation measures, such as:		
a. Storing hay bales and equipment in areas less likely to be flooded?	☐ Yes	☐ No
b. Installing ponds or swales to capture stormwater?	☐ Yes	☐ No
c. Planting vegetation that can tolerate inundation?	☐ Yes	☐ No
d. Using land management practices to improve the capability of the soil on their lands to retain water?	☐ Yes	☐ No
3. Has the community adopted flood plain development limits that go beyond FEMA's minimum standards for Special Flood Hazard Areas and also prohibit or reduce any new encroachment and fill in river corridors and Fluvial Erosion Hazard areas?	☐ Yes	☐ No
4. Has the community implemented development regulations that incorporate approaches and standards to protect land in vulnerable areas, including:		
a. Fluvial erosion hazard zoning?	☐ Yes	☐ No
b. Agricultural or open space zoning?	☐ Yes	☐ No
c. Conservation or cluster subdivision ordinances, where appropriate?	☐ Yes	☐ No
d. Other zoning or regulatory tools that limit development in areas subject to flooding, including river corridors and Special Flood Hazard Areas?	☐ Yes	☐ No
Protect People, Buildings, and Facilities in Vulnerable Settlements (Learn more in Section 3.B, pp. 19-26)		
1. Do the local comprehensive plan and Hazard Mitigation Plan identify developed areas that have been or are likely to be flooded?	☐ Yes	☐ No
a. If so, does the comprehensive plan discourage development in those areas or require strategies to reduce damage to buildings during floods (such as elevating heating, ventilation, and air conditioning (HVAC) systems and flood-proofing basements)?	☐ Yes	☐ No
b. Does the Hazard Mitigation Plan identify critical facilities and infrastructure that are located in vulnerable areas and should be protected, repaired, or relocated (e.g., town facilities, bridges, roads, and wastewater facilities)?	☐ Yes	☐ No
2. Do land development regulations and building codes promote safer building and rebuilding in flood-prone areas? Specifically:		

FLOOD RESILIENCE CHECKLIST		
a. Do zoning or flood plain regulations require elevation of two or more feet above base flood elevation?	☐ Yes	☐ No
b. Does the community have the ability to establish a temporary post-disaster building moratorium on all new development?	☐ Yes	☐ No
c. Have non-conforming use and structure standards been revised to encourage safer rebuilding in flood-prone areas?	☐ Yes	☐ No
d. Has the community adopted the International Building Code or American Society of Civil Engineers (ASCE) standards that promote flood-resistant building?	☐ Yes	☐ No
e. Does the community plan for costs associated with follow-up inspection and enforcement of land development regulations and building codes?	☐ Yes	☐ No
3. Does the community require developers who are rebuilding in flood-prone locations to add additional flood storage capacity in any new redevelopment projects such as adding new parks and open space and allowing space along the river's edge for the river to move during high-water events?	☐ Yes	☐ No
4. Is the community planning for development (e.g., parks, river-based recreation) along the river's edge that will help connect people to the river AND accommodate water during floods?	☐ Yes	☐ No
5. Does the comprehensive plan or Hazard Mitigation Plan discuss strategies to determine whether to relocate structures that have been repeatedly flooded, including identifying an equitable approach for community involvement in relocation decisions and potential funding sources (e.g., funds from FEMA, stormwater utility, or special assessment district)?	☐ Yes	☐ No
Plan for and Encourage New Development in Safer Areas (Learn more in Section 3.C, pp. 26-27)		
1. Does the local comprehensive plan or Hazard Mitigation Plan clearly identify safer growth areas in the community?	☐ Yes	☐ No
2. Has the community adopted policies to encourage development in these areas?	☐ Yes	☐ No
3. Has the community planned for new development in safer areas to ensure that it is compact, walkable, and has a variety of uses?	☐ Yes	☐ No
4. Has the community changed their land use codes and regulations to allow for this type of development?	☐ Yes	☐ No
5. Have land development regulations been audited to ensure that development in safer areas meets the community's needs for off-street parking requirements, building height and density, front-	☐ Yes	☐ No

FLOOD RESILIENCE CHECKLIST

yard setbacks and that these regulations do not unintentionally inhibit development in these areas?		
6. Do capital improvement plans and budgets support development in preferred safer growth areas (e.g., through investment in wastewater treatment facilities and roads)?	☐ Yes	☐ No
7. Have building codes been upgraded to promote more flood-resistant building in safer locations?	☐ Yes	☐ No

Implement Stormwater Management Techniques throughout the Whole Watershed

(Learn more in Section 3.D, pp. 27-31)

1. Has the community coordinated with neighboring jurisdictions to explore a watershed-wide approach to stormwater management?	☐ Yes	☐ No
2. Has the community developed a stormwater utility to serve as a funding source for stormwater management activities?	☐ Yes	☐ No
3. Has the community implemented strategies to reduce stormwater runoff from roads, driveways, and parking lots?	☐ Yes	☐ No
4. Do stormwater management regulations apply to areas beyond those that are regulated by federal or state stormwater regulations?	☐ Yes	☐ No
5. Do stormwater management regulations encourage the use of green infrastructure techniques?	☐ Yes	☐ No
6. Has the community adopted tree protection measures?	☐ Yes	☐ No
7. Has the community adopted steep slope development regulations?	☐ Yes	☐ No
8. Has the community adopted riparian and wetland buffer requirements?	☐ Yes	☐ No

Appendix D: Flood Resilience Resources

The following resources, many of which are discussed in this report, might be helpful as your community assesses its flood resilience and begins implementing the strategies described in this report. The resources are organized according to the sections of the report:

- Overall Strategies to Enhance Flood Resilience

- River Corridors: Conserve Land and Discourage Development

- Vulnerable Settlements: Protect People, Buildings, and Facilities

- Safer Areas: Plan for New Development

- The Whole Watershed: Manage Stormwater

- State Policy Resources

- Selected Federal Resources

Overall Strategies to Enhance Flood Resilience

Smart Growth and Flood Resilience Checklists and Resources

Coastal Resilience Index: A Community Self-Assessment. Mississippi-Alabama Sea Grant Consortium and National Oceanic and Atmospheric Administration. 2010. http://www.southernclimate.org/documents/resources/Coastal_Resilience_Index_Sea_Grant.pdf.

Essential Smart Growth Fixes for Rural Planning, Zoning, and Development Codes. U.S. Environmental Protection Agency (EPA). 2012. http://www.epa.gov/smartgrowth/essential_fixes.htm#part2.

Preparing for the Next Flood: Vermont Floodplain Management. Land Use Institute, Vermont Law School. 2009. http://www.vermontlaw.edu/Documents/VLS.065.09%20LAND%20USE%20PAPER_PFF.pdf.

Resilient Communities Scorecard: A Tool for Assessing Your Community. Vermont Natural Resources Council. 2013. http://vnrc.org/resources/community-planning-toolbox/tools/vermont-smart-growth-score-card.

Smart Growth for Coastal and Waterfront Communities. National Oceanic and Atmospheric Administration, U.S. EPA, ICMA, and Sea Grant Rhode Island. 2009. http://coastalsmartgrowth.noaa.gov/smartgrowth_fullreport.pdf.

Smart Growth Implementation Toolkit. Smart Growth America. http://www.smartgrowthamerica.org/leadership-institute/implementation-tools.

Integrating Hazard Mitigation Plans and Comprehensive Plans

Hazard Mitigation Planning: Integrating Best Practices into Planning. American Planning Association, Planning Advisory Service. 2010. http://www.fema.gov/media-library-data/20130726-1739-25045-4373/pas_560_final.pdf.

Integrating Hazard Mitigation Into Local Planning: Case Studies and Tools for Community Officials. Federal Emergency Management Agency (FEMA). 2013. http://www.fema.gov/media-library/assets/documents/31372?id=7130.

National Flood Insurance Program Community Rating System

Changes to the Community Rating System to Improve Disaster Resiliency and Community Sustainability. FEMA. 2013. http://www.fema.gov/media-library-data/20130726-1907-25045-6528/changes_to_crs_system_2013.pdf.

Community Rating System. FEMA. 2012. http://www.fema.gov/media-library-data/20130726-1605-20490-0645/communityratingsystem_2012.pdf

River Corridors: Conserve Land and Discourage Development

Land Acquisition/Buyouts

Charlotte Land Trust. "Process of Conservation." http://www.charlottelandtrust.org/conservation/.

FEMA. "Hazard Mitigation Assistance – Property Acquisition (Buyouts)." http://www.fema.gov/application-development-process/hazard-mitigation-assistance-property-acquisition-buyouts.

Napa County, CA. "Flood Control and Water Conservation District." http://www.countyofnapa.org/Pages/DepartmentContent.aspx?id=4294971816.

Property Acquisition Handbook for Local Communities: A Summary for States. FEMA. 1998. http://www.fema.gov/media-library/assets/documents/3117.

Town of Charlotte, VT. *Selectboard Meeting Notes.* August 9, 2010. http://www.charlottevt.org/vertical/sites/%7B5618C1B5-BAB5-4588-B4CF-330F32AA3E59%7D/uploads/%7B56035679-59BD-419A-8D4E-7D23A177FD50%7D.PDF.

Transfer of Development Rights

Maryland's Transfer of Development Rights Programs. 2010. http://www.cga.ct.gov/2010/rpt/2010-R-0464.htm.

The New Jersey Pinelands Development Credit (PDC) Program. New Jersey Pinelands Commission. 2012. http://www.nj.gov/pinelands/infor/fact/PDCfacts.pdf.

TDR Program Overview. Department of Economic Development, Agricultural Services Division. 2006. http://www6.montgomerycountymd.gov/content/ded/agservices/pdffiles/tdr_info.pdf.

Tax Strategies: Sales Taxes, Tax Credits, and Current Use Taxation

Commonwealth of Virginia. *Riparian forest buffer protection for waterways tax credit.* Section 58.1-339.10. http://leg1.state.va.us/cgi-bin/legp504.exe?000+cod+58.1-339.10.

Sales Tax: Earmarked for Open Space. University of Washington. 2005. http://depts.washington.edu/open2100/pdf/3_OpenSpaceImplement/Implementation_Mechanisms/sales_tax.pdf.

Vermont Department of Taxes. "Current Use." http://www.state.vt.us/tax/pvrcurrentuse.shtml.

Disaster Mitigation for Agricultural and Other Landowners

Extension Disaster Education Network. "Extension Disaster Education Network." http://eden.lsu.edu/Pages/default.aspx.

National Voluntary Organizations Active in Disaster. "National Voluntary Organizations Active in Disaster." http://www.nvoad.org/.

Ready Ag: Disaster and Defense Preparedness for Production Agriculture. Penn State Cooperative Extension. 2010. http://readyag.psu.edu/pdfs/ReadyAG_DAIRYandGENERALWorkbook.pdf.

Small Business Technology Development Center. "Small Business Technology Development Center." http://www.sbtdc.org.

Special Considerations for Agricultural Producers-Preparing for a Flood or a Flash Flood. University of Florida Cooperative Extension Service. 1998. http://disaster.ifas.ufl.edu/PDFS/CHAP09/D09-07.pdf.

Texas A&M AgriLife Extension. "Texas Extension Disaster Education Network." http://texashelp.tamu.edu.

No Adverse Impact Flood Plain Management

Association of State Flood Plain Managers. "No Adverse Impact Floodplain Management." http://www.floods.org/index.asp?menuID=349&fir.

No Adverse Impact Floodplain Management Community Case Studies. Association of State Flood Plain Managers. 2004. http://www.floods.org/PDF/NAI_Case_Studies.pdf.

No Adverse Impact Status Report: Helping Communities Implement NAI. Association of State Flood Plain Managers. 2002. http://www.floods.org/NoAdverseImpact/NAI_Status_Report.pdf.

Fluvial Erosion Hazard Zoning

Municipal Guide to Fluvial Erosion Hazard Mitigation. Vermont Agency of Natural Resources. 2011. http://www.vtwaterquality.org/rivers/docs/rv_municipalguide.pdf.

Town of Bennington, VT. *Fluvial Erosion Hazard Overlay District*. Adopted Apr., 27, 2009. http://www.benningtonplanningandpermits.com/BPC/wp-content/uploads/2011/02/fehr.pdf.

Agricultural/Open Space Zoning

Blaine County, ID. *County Code, Title 9, Chapter 6B: Resource Conservation District (RC-160)*. Passed Mar. 19, 2013. http://www.sterlingcodifiers.com/codebook/index.php?book_id=450.

Minnesota Department of Agriculture. "Minnesota's Agricultural Land Preservation Statutes." http://www.mda.state.mn.us/protecting/conservation/preservation/statutes.aspx.

Stearns County, MN. "Zoning Districts." http://www.co.stearns.mn.us/Environment/LandUseandSubdivision/Zoning/ZoningDistricts.

Town of Colchester, VT. *Zoning Regulations, Table A-2 Dimensional Standards*. Amended Dec. 10, 2013. http://colchestervt.gov/PlanningZ/regs/Zoning/A-2-DimensionalReq.pdf.

Conservation/Cluster Subdivision Ordinances

Town of Hartford, VT. *Ordinances*. Jul. 25, 2013. https://law.resource.org/pub/us/code/city/vt/Hartford.html#13455808.

Town of St. George, VT. *Land Use Regulations*. Adopted Jul. 22, 2010. http://www.stgeorgevt.com/pdfs/Regulations%20and%20Bylaws/SGLUR(22Jul2010)lr.pdf.

Town of Windsor, VT. *Subdivision Regulations*. Amended Sept. 26, 2006. http://swcrpc.org/wp-content/uploads/2013/11/Windsor-Subdivision-Regulations-2006.pdf.

Town of Windsor, VT. *Zoning Regulations*. Amended Sept. 25, 2007. http://swcrpc.org/wp-content/uploads/2013/11/Windsor-Zoning-Regulations-2007.pdf.

Vulnerable Settlements: Protect People, Buildings, and Facilities

Streambank Stabilization

Engineering with Nature: Alternative Techniques to Riprap Bank Stabilization. FEMA. http://www.fema.gov/pdf/about/regions/regionx/Engineering_With_Nature_Web.pdf.

Elevating Above Base Flood Elevation

Innovative Land Use Planning Techniques: A Handbook for Sustainable Development. Chapter 2.7 Flood Hazard Area Zoning. New Hampshire Department of Environmental Services. 2008. http://des.nh.gov/organization/divisions/water/wmb/repp/documents/ilupt_complete_handbook.pdf.

No Adverse Impact Floodplain Management Community Case Studies. Association of State Flood Plain Managers. 2004. http://www.floods.org/PDF/NAI_Case_Studies.pdf.

No Adverse Impact Status Report: Helping Communities Implement NAI. Association of State Flood Plain Managers. 2002. http://www.floods.org/NoAdverseImpact/NAI_Status_Report.pdf.

Building Code Upgrades

Code Information Sheet: Permit and Licensing Requirements. Vermont Department of Public Safety. http://firesafety.vermont.gov/sites/firesafety/files/pdf/Code%20Info%20Sheets/2012%20permit%20requirements.pdf.

FEMA. "Highlights of ASCE 24-05, Flood Resistant Design and Construction (2010)." http://www.fema.gov/library/viewRecord.do?id=3515.

International Code Council. "International Code Council." http://www.iccsafe.org.

International Code Council. "International Green Construction Code." http://www.iccsafe.org/cs/IGCC.

Vermont Department of Public Safety. "Code Information Sheets." http://firesafety.vermont.gov/resources/code_sheets.

Safer Areas: Plan for New Development

Identifying Safer Locations for Development in Vermont

Vermont Agency of Natural Resources. "Flood Resilience." https://outside.vermont.gov/agency/ANR/FloodResilience/Pages/default.aspx.

The Whole Watershed: Manage Stormwater

Watershed-Wide Approaches

Chittenden County, VT. "Smarter WaterWays." http://www.smartwaterways.org/.

Mad River Valley Erosion Study Final Report. Watershed Consulting Associates, LLC. 2012. http://friendsofthemadriver.org/documents/MRV_Road_Erosion_Study_Report.pdf.

Metropolitan North Georgia Water Planning District. "Watershed Management Plan." http://www.northgeorgiawater.com/plans/watershed-management-plan.

Stormwater Management Regulation in the Mad River Valley: Review and Recommendations. Watershed Consulting Associates, LLC. 2013. http://www.friendsofthemadriver.org/documents/MRVStormwater_Scoping_Study_Spring_2013_.pdf.

Stormwater Utilities

Funding Stormwater Programs. U.S. EPA. 2009. http://www.epa.gov/region1/npdes/stormwater/assets/pdfs/FundingStormwater.pdf.

Managing Roads, Driveways, and Parking Lots

Guidelines for the Design of Stream/Road Crossings for Passage of Aquatic Organisms in Vermont. Vermont Department of Fish and Wildlife. 2009. http://www.vtfishandwildlife.com/library/Reports_and_Documents/Aquatic%20Organism%20Passage%20at%20Stream%20Crossings/_Guidelines%20for%20the%20Design%20of%20Stream_Road%20Crossings%20for%20Passage%20of%20Aquatic%20Organisms%20in%20Vermont.pdf.

Vermont Town Road and Bridge Standards, Culverts and Bridges. 2013. http://www.vlct.org/assets/News/Current/Town_Road_Bridge_Standards.pdf.

Using Green Infrastructure in Stormwater Regulations

Stormwater Management Plan. City of South Burlington, VT. 2013.
http://www.watershedmanagement.vt.gov/stormwater/docs/ms4/sw_SBurlington_SWMP.pdf.

Stormwater Management Plan. Town of Rutland, VT. 2013.
http://www.vtwaterquality.org/stormwater/docs/ms4/sw_TownofRutland_MS4_SWMP.pdf.

Town of Williston, VT. *Unified Development Bylaw.* Chapter 29, Watershed Health. Amended Jul. 19, 2010.
http://town.williston.vt.us/vertical/Sites/%7BF506B13C-605B-4878-8062-
87E5927E49F0%7D/uploads/%7B2920AC61-60E4-483B-8A02-015028396045%7D.PDF.

U.S. EPA. "Stormwater Management Best Practices."
http://www.epa.gov/oaintrnt/stormwater/best_practices.htm.

U.S. EPA. "Why Green Infrastructure?" http://water.epa.gov/infrastructure/greeninfrastructure/gi_why.cfm.

Williston, VT. "Stormwater." http://www.town.williston.vt.us/index.asp?Type=B_BASIC&SEC=%7BACC6B21E-
0FDB-497F-8A5A-62CDFF871272%7D.

Tree Protection

Folly Beach, SC. *Code of Ordinances.* Title XV, Section 166.01, Tree Protection. Passed Jul. 23, 2013.
http://www.amlegal.com/nxt/gateway.dll/South%20Carolina/follybeach/follybeachsouthcarolinacodeofordinance
s?f=templates$fn=default.htm$3.0$vid=amlegal:follybeach_sc.

Town of Wellesley MA. *Rules and Regulations Relative to the Administration of Section XVIE: Tree Preservation and Protection.* http://www.wellesleyma.gov/Pages/WellesleyMA_Planning/TreeBylawRulesRegs6.27.11.pdf.

Steep Slope Development Regulations

Salt Lake County, UT. *Code of Ordinances.* Title 19: Zoning, Chapter 19.72: Foothills and Canyons Overlay Zone.
http://library.municode.com/HTML/16602/level2/TIT19ZO_CH19.72FOCAOVZO.html.

Town of Williston, VT. *Unified Development Bylaw.* Chapter 29, Watershed Health. Amended Jul. 19, 2010.
http://town.williston.vt.us/vertical/Sites/%7BF506B13C-605B-4878-8062-
87E5927E49F0%7D/uploads/%7B2920AC61-60E4-483B-8A02-015028396045%7D.PDF.

Stream and Wetland Buffer Regulations

Mitchell, Paul. *The Scientific Justification for Stream Buffers.* University of Georgia Land Use Clinic. 2006.
http://www.rivercenter.uga.edu/publications/pdf/luc_buffer_fact_sheet.pdf.

State Policy Resources

Background/Overview of State Policy Issues

Smith, Gavin. *Planning for Post-Disaster Recovery: A Review of the United States Disaster Assistance Framework.* Gavin Smith. Island Press, 2012.

Smith, Gavin, and Dylan Sandler. *State Disaster Recovery Planning Guide.* U.S. Department of Homeland Security Coastal Hazards Center of Excellence, University of North Carolina at Chapel Hill. 2012.
http://coastalhazardscenter.org/dev/wp-content/uploads/2012/05/State-Disaster-Recovery-Planning-
Guide_2012.pdf.

Smith, Gavin, Dylan Sandler, and Mikey Goralnik. "Assessing State Policy Linking Disaster Recovery, Smart Growth, and Resilience in Vermont Following Tropical Storm Irene." *Vermont Journal of Environmental Law.* Vol. 15 (2013). 66-102. http://vjel.vermontlaw.edu/files/2013/11/Smith.pdf.

Smith, Gavin, Dylan Sandler, and Mikey Goralnik. *Vermont State Agency Policy Options: Smart Growth Implementation Assistance Program, Disaster Recovery and Long-Term Resilience Planning in Vermont.* U.S.

Department of Homeland Security Coastal Hazards Center of Excellence, University of North Carolina at Chapel Hill. http://accd.vermont.gov/sites/accd/files/Documents/strongcommunities/cpr/VT-StateAgencyPolicyOptionsFINAL_web.pdf.

State-Level Initiatives and Resources

California Emergency Management Agency. "Local Hazard Mitigation Planning Program (LHMP)." http://hazardmitigation.calema.ca.gov/plan/local_hazard_mitigation_plan_lhmp.

California Governor's Office of Emergency Services. "Hazard Mitigation." http://www.calema.ca.gov/hazardmitigation.

Culvert Inventory and Inspection Manual. New York State Department of Transportation. 2006. https://www.dot.ny.gov/divisions/operating/oom/transportation-maintenance/repository/CulvertInventoryInspectionManual.pdf.

Culvert Management Manual. Ohio Department of Transportation. 2014. http://www.dot.state.oh.us/Divisions/Engineering/Hydraulics/Culvert%20Management/Culvert%20Management%20Manual/CMM%20-%20January2014.pdf.

Handbook of Emergency Management for State-Level Transportation Agencies. San Jose State University. 2010. http://transweb.sjsu.edu/MTIportal/research/publications/documents/COOP%20COG%20I_Vince_022410.pdf.

Louisiana Recovery Authority Strategic Plan: FY 2008/2009. Louisiana Recovery Authority. http://lra.louisiana.gov/assets/docs/searchable/StrategicPlan0809.pdf.

University of Iowa School of Urban and Regional Planning. "RIO Iowa Project." http://rio.urban.uiowa.edu/.

Wisconsin Department of Military Affairs, Division of Emergency Management. "2012 All-Hazards Mitigation Planning Workshop Presentations and Handouts." http://emergencymanagement.wi.gov/mitigation/Mitigation_Workshop/toc.asp.

State Statutes for Integrating Flood Resilience into Comprehensive Plans

State of Rhode Island. *General Laws*. Title 45: Towns and Cities, Chapter 45-22.2: Rhode Island Comprehensive Planning and Land Use Act, Section 45-22.2-6: Required Content of a Comprehensive Plan. http://webserver.rilin.state.ri.us/Statutes/TITLE45/45-22.2/45-22.2-6.HTM.

State of Vermont. *Vermont Statutes*. Title 24: Municipal and County Government, Chapter 117: Municipal and Regional Planning and Development, Sub-Chapter 5: Municipal Development Plan, Section 4382: The Plan for a Municipality. http://www.leg.state.vt.us/statutes/fullchapter.cfm?Title=24&Chapter=117.

Selected Federal Resources

Federal Emergency Management Agency

Changes to the Community Rating System to Improve Disaster Resiliency and Community Sustainability. FEMA. 2013. http://www.fema.gov/media-library-data/20130726-1907-25045-6528/changes_to_crs_system_2013.pdf.

FEMA. "Community Emergency Response Teams." https://www.fema.gov/community-emergency-response-teams.

Community Rating System. FEMA. 2012. http://www.fema.gov/media-library-data/20130726-1605-20490-0645/communityratingsystem_2012.pdf.

Community Rating System Communities by State. FEMA. 2012. http://www.fema.gov/media-library-data/20130726-1830-25045-0453/crosstab_bystate_4may_2012.pdf.

FEMA. "Community Planning and Capacity Building." http://www.fema.gov/community-planning-and-capacity-building.

FEMA. "Flood Insurance Reform." http://www.fema.gov/flood-insurance-reform.

National Flood Insurance Program. "Floodsmart." https://www.floodsmart.gov/floodsmart/.

FEMA. "Hazard Mitigation Assistance – Property Acquisition (Buyouts)." http://www.fema.gov/application-development-process/hazard-mitigation-assistance-property-acquisition-buyouts.

FEMA. "Hazard Mitigation Grant Program." http://www.fema.gov/hazard-mitigation-grant-program.

FEMA. "Multi-Hazard Mitigation Planning." http://www.fema.gov/multi-hazard-mitigation-planning.

FEMA. "National Disaster Recovery Framework." http://www.fema.gov/national-disaster-recovery-framework.

FEMA. "Pre-Disaster Mitigation Grant Program." http://www.fema.gov/pre-disaster-mitigation-grant-program.

Property Acquisition Handbook for Local Communities: A Summary for States. FEMA. 1998. http://www.fema.gov/media-library/assets/documents/3117.

FEMA. "Public Assistance: Local, State, Tribal and Non-Profit." http://www.fema.gov/public-assistance-local-state-tribal-and-non-profit.

Reducing Damage from Localized Flooding: A Guide for Communities. http://www.fema.gov/media-library/assets/documents/1012?id=1448.

FEMA. "Response and Recovery." http://www.fema.gov/response-recovery.

U.S. Department of Agriculture

U.S. Department of Agriculture (USDA). "Conservation." http://www.usda.gov/wps/portal/usda/usdahome?navid=CONSERVATION.

USDA, Forest Service. "Conservation Reserve Enhancement Program – Vermont." http://www.fsa.usda.gov/FSA/newsReleases?area=newsroom&subject=landing&topic=pfs&newstype=prfactsheet&type=detail&item=pf_20110214_consv_en_crepvt01.html.

National Institute of Food and Agriculture. "Cooperative Extension Offices." http://www.csrees.usda.gov/Extension/.

USDA. "USDA Rural Development Programs." http://www.rurdev.usda.gov/programsandopportunities.html.

U.S. Department of Housing and Urban Development

Vermont Agency of Commerce and Community Development. "Community Development Block Grant Disaster Recovery Funds." http://accd.vermont.gov/strong_communities/opportunities/funding/cdbgdr.

U.S. Department of Housing and Urban Development (HUD). "Community Development Block Grant Entitlement Communities Grants." http://portal.hud.gov/hudportal/HUD?src=/program_offices/comm_planning/communitydevelopment/programs/entitlement#eligibleactivities.

HUD. "Community Development Block Grant Program – CDBG." http://portal.hud.gov/hudportal/HUD?src=/program_offices/comm_planning/communitydevelopment/programs.

HUD. "Pre-Disaster Planning for Permanent Housing Recovery." http://www.huduser.org/portal/publications/pre_disasterplanning.html.

HUD. "State Administered Community Development Block Grant." http://portal.hud.gov/hudportal/HUD?src=/program_offices/comm_planning/communitydevelopment/programs/stateadmin.

U.S. Department of Transportation

Emergency Relief Manual. Federal Highway Administration. 2013. http://www.fhwa.dot.gov/reports/erm/er.pdf.

U.S. Environmental Protection Agency

U.S. EPA. "Adaptation Strategies Guide for Water Utilities." http://water.epa.gov/infrastructure/watersecurity/climate/upload/epa817k11003.pdf.

U.S. EPA. "BASINS 4 Climate Assessment Tool." http://cfpub.epa.gov/ncea/global/recordisplay.cfm?deid=203460.

U.S. EPA. "Climate Ready Water Utilities." http://water.epa.gov/infrastructure/watersecurity/climate/index.cfm.

U.S. EPA. "Climate Resilience Evaluation and Awareness Tool." http://water.epa.gov/infrastructure/watersecurity/climate/creat.cfm.

U.S. EPA. "National Stormwater Calculator." http://epa.gov/nrmrl/wswrd/wq/models/swc/.

U.S. EPA. "Smart Growth Program." http://www.epa.gov/smartgrowth.

U.S. EPA. "Smart Growth Program Climate Page." http://www.epa.gov/smartgrowth/climatechange.htm.

U.S. EPA. "Smart Growth Technical Assistance in Iowa." http://www.epa.gov/smartgrowth/iowa_techasst.htm.

U.S. EPA. "Stormwater Management Best Practices." http://www.epa.gov/oaintrnt/stormwater/best_practices.htm.

U.S. EPA. "Vulnerability Self Assessment Tool." http://water.epa.gov/infrastructure/watersecurity/techtools/vsat.cfm.

U.S. EPA. "Why Green Infrastructure?" http://water.epa.gov/infrastructure/greeninfrastructure/gi_why.cfm.